Diesel Engine - Combustion, Emissions and Condition Monitoring

Diesel Engine - Combustion, Emissions and Condition Monitoring

Editor
Thanh Phuc Van

Diesel Engine - Combustion, Emissions and Condition Monitoring
Edited by **Thanh Phuc Van**

ISBN: 978-1-68117-474-7
Library of Congress Control Number: 2016942832

© 2017 by
Scitus Academics LLC,
616, Corporate Way, Suite 2, 4766,
Valley Cottage, NY 10989

This book contains information obtained from highly regarded resources. Copyright for individual articles remains with the authors as indicated. All chapters are distributed under the terms of the Creative Commons Attribution License, which permits unrestricted use, distribution, and reproduction in any medium, provided the original author and source are credited.

Notice

Reasonable efforts have been made to publish reliable data and views articulated in the chapters are those of the individual contributors, and not necessarily those of the editors or publishers. Editors or publishers are not responsible for the accuracy of the information in the published chapters or consequences of their use. The publisher believes no responsibility for any damage or grievance to the persons or property arising out of the use of any materials, instructions, methods or thoughts in the book. The editors and the publisher have attempted to trace the copyright holders of all material reproduced in this publication and apologize to copyright holders if permission has not been obtained. If any copyright holder has not been acknowledged, please write to us so we may rectify.

Printed in India by Replika Press

Preface

Diesel engines, also known as CI engines, possess a wide field of applications as energy converters because of their higher efficiency. However, diesel engines are a major source of NOX and particulate matter (PM) emissions.Like a gasoline engine, a diesel engine is a type of internal combustion engine. Combustion is another word for burning, and internal means inside, so an internal combustion engine is simply one where the fuel is burned inside the main part of the engine (the cylinders) where power is produced. That's very different from an external combustion engine such as those used by old-fashioned steam locomotives. The diesel engine has the highest thermal efficiency (engine efficiency) of any practical internal or external combustion engine due to its very high expansion ratio and inherent lean burn which enables heat dissipation by the excess air. A small efficiency loss is also avoided compared to two-stroke non-direct-injection gasoline engines since unburnt fuel is not present at valve overlap and therefore no fuel goes directly from the intake/injection to the exhaust. Low-speed diesel engines (as used in ships and other applications where overall engine weight is relatively unimportant) can have a thermal efficiency that exceeds 50%. We are currently experiencing an oil crisis world-wide. Gaseous fuels like natural gas, pure hydrogen gas, biomass-based and coke-based syngas can be considered as alternative fuels for diesel engines.

Diesel Engine - Combustion, Emissions and Condition Monitoring describescombustion and exhaust emissions features. Reliable early detection of malfunction and failure of any parts in diesel engines can save the engine from failing completely and protect high repair cost. Tools are discussed in this book to discover common failure approaches of diesel engine that can identify early signs of failure.

Table of Contents

Chapter 1	The Effect of Split Injection on the Combustion and Emissions in DI and IDI Diesel Engines	1
Chapter 2	Analytical Methodologies for the Control of Particle-Phase Polycyclic Aromatic Compounds from Diesel Engine Exhaust	37
Chapter 3	A Model-Free Diagnosis Approach for Intake Leakage Detection and Characterization in Diesel Engines	71
Chapter 4	Homogeneous Charge Compression Ignition Combustion: Challenges and Proposed Solutions	95
Chapter 5	NO_x Storage and Reduction forDiesel Engine Exhaust Aftertreatment	129
Chapter 6	Analysis of Two Stroke Marine Diesel Engine Operation Including Turbocharger Cut-Out by Using a Zero-Dimensional Model	173
Chapter 7	Performance Analysis of the Vehicle Diesel Engine-ORC Combined System Based on a Screw Expander	209

Chapter 8   Combustion and Exhaust Emission Characteristics of Diesel Micro-Pilot Ignited Dual-Fuel Engine   237

Chapter 9   Optimization of Diesel Engine with Dual-Loop EGR by Using DOE Method   277

Index   297

CHAPTER 1

The Effect of Split Injection on the Combustion and Emissions in DI and IDI Diesel Engines

S. Jafarmadar

[1] Mechanical Engineering Department, Technical Education Faculty, Urmia University, Urmia, West Azerbaijan, Iran

1. INTRODUCTION

The major pollutants from diesel engines are NOx and soot. NOx and soot emissions are of concerns to the international community. They have been judged to pose a lung cancer hazard for humans as well as elevating the risk of non-cancer respiratory ailments. These emissions react in the atmosphere in the presence of sunlight to form ground-level ozone. Ground-level ozone is a major component of smog in our cities and in many rural areas as well. In addition, NOx reacts with water, oxygen and oxidants in the atmosphere to form acid rain. Furthermore, the indirect effect of NOx emission to global warming should be noted. It is possible that NOx emission causes an increase secondary emission formation and global warming.

Stringent exhaust emission standards require the simultaneous reduction of soot and NOx for diesel engines, however it seems to be very difficult to reduce NOx emission without increasing soot emission by injection timing. The reason is that there always is a contradiction between NOx and soot emissions when the injection timing is retarded or advanced.

Split injection has been shown to be a powerful tool to simultaneously reduce soot and NOx emissions for DI and IDI diesel engines when the injection timing is optimized. It is defined as splitting the main single

1. Introduction

injection profile in two or more injection pulses with definite delay dwell between the injections. However, an optimum injection scheme of split injection for DI and IDI diesel engines has been always under investigation.

Generally, the exhaust of IDI diesel engines because of high turbulence intensity is less smoky when compared to DI diesel engines [1]. Hence, investigation the effect of split injection on combustion process and pollution reduction of IDI diesel engines can be quite valuable.

In an IDI diesel engine, the combustion chamber is divided into the pre-chamber and the main chamber, which are linked by a throat. The pre-chamber approximately contains 50 % of the combustion volume when the piston is at TDC. This geometrical represents an additional difficulty to those deals with in the DI combustion chambers. Fuel injects into the pre-combustion chamber and air is pushed through the narrow passage during the compression stroke and becomes turbulent within the pre-chamber. This narrow passage speeds up the expanding gases more.

In the recent years, the main studies about the effect of the split injection on the combustion process and pollution of DI and IDI diesel engines are as follows.

Bianchi et al [2] investigated the capability of split injection in reducing NOx and soot emissions of HSDI Diesel engines by CFD code KIVA-III. Computational results indicate that split injection is very effective in reducing NOx, while soot reduction is related to a better use of the oxygen available in the combustion chamber.

Seshasai Srinivasan et al [3] studied the impact of combustion variations such as EGR (exhaust gas recirculation) and split injection in a turbo-charged DI diesel engine by an Adaptive Gradient-Based Algorithm. The predicted values by the modeling, showed a good agreement with the experimental data. The best case showed that the nitric oxide and the particulates could be reduced by over 83 % and almost 24 %, respectively while maintaining a reasonable value of specfic fuel consumption.

Shayler and Ng [4] used the KIVA-III to investigate the influence of mass ratio of two plus injections and delay dwell on NOx and soot emissions. Numerical conclusions showed that when delay dwell is small, soot is lowered but NOx is increased. In addition, when delay dwell is large, the second injection has very little influence on soot production and oxidation associated with the first injection.

Chryssakis et al [5] studied the effect of multiple injections on combustion process and emissions of a DI diesel engine by using the multidimensional code KIVAIII. The results indicated that employing a post-injection combined with a pilot injection results in reduced soot formation; while the NOx concentration is maintained at low levels.

Lechner et al [6] analyzed the effect of spray cone angle and advanced injection-timing strategy to achieve partially premixed compression ignition (PCI) in a DI diesel engine. The authors proved that low flow rate of the fuel; 60-degree spray cone angle injector strategy, optimized EGR and split injection strategy could reduce the engine NOx emission by 82 % and particular matter by 39 %.

Ehleskog [7] investigated the effect of split injection on the emission formation and engine performance of a heavy-duty DI diesel engine by KIVA-III code. The results revealed that reductions in NOx emissions and brake-specific fuel consumption were achieved for short dwell times whereas they both were increased when the dwell time was prolonged.

Sun and Reitz [8] studied the combustion and emission of a heavy-duty DI diesel engine by multi-dimensional Computational Fluid Dynamics (CFD) code with detailed chemistry, the KIVA-CHEMKIN. The results showed that the start of late injection timing in two-stage combustion in combination with late IVC timing and medium EGR level was able to achieve low engine-out emissions.

Verbiezen et al [9] investigated the effect of injection timing and split injection on NOx concentration in a DI diesel engine experimentally. The results showed that advancing the injection timing causes NOx increase. Also, maximum rate of heat release is significantly reduced by the split injection. Hence, NOx is reduced significantly.

Abdullah et al [10] progressed an experimental research for optimizing the variation of multiple injections on the engine performance and emissions of a DI diesel engine. The results show that, the combination of high pressure multiple injections with cooled EGR produces better overall results than the combination of low injection pressure multiple injections without EGR.

Jafarmadar and Zehni [11] studied the effect of split injection on combustion and pollution of a DI diesel engine by Computational Fluid Dynamics (CFD) code. The results show that 25 % of total fuel injected in the second pulse, reduces the total soot and NOx emissions effectively in DI diesel engines. In addition, the optimum delay dwell between the two injection pulses was about 25ºCA.

Showry and Rajo [12] carried out the effect of triple injection on combustion and pollution of a DI diesel engine by FLUENT CFD code and concluded that 10° is an optimum delay between the injection pulses for triple injection strategy. The showed that split injections take care of reducing of PM without increasing of NOx level.

As mentioned, the effect of split injection on combustion and emission of DI Diesel engines has been widely studied up to now. However, for IDI diesel engines, the study of split injection strategy in order to reduce emissions is confined to the research of Iwazaki et al [13] that investigated the effects of early stage injection and two-stage injection on the combustion characteristics of an IDI diesel engine experimentally. The results indicated that NOx and smoke emissions are improved by two-stage injection when the amount of fuel in the first injection was small and the first injection timing was advanced from -80 to -100° TDC.

At the present work, the effect of the split injection on combustion and pollution of DI and IDI diesel engines is studied at full load state by the CFD code FIRE. The target is to obtain the optimum split injection case in which the total exhaust NOx and soot concentrations are more reduced than the other cases. Three different split injection schemes, in which 10, 20 and 25 % of total fuel injected in the second pulse, have been considered. The delay dwell between injections is varied from 5°CA to 30°CA with the interval 5°CA.

2. INITIAL AND BOUNDRY CONDITIONS

Calculations are carried out on the closed system from Intake Valve Closure (IVC) at 165°CA BTDC to Exhaust Valve Open (EVO) at 180°CA ATDC. Fig. 1-a and Fig. 1-b show the numerical grid, which is designed to model the geometry of combustion chamber of IDI engine and contains a maximum of 42200 cells at BTDC. As can be seen from the figure Fig. 1-c, Grid dependency is based on the in-cylinder pressure and present resolution is found to give adequately grid independent results. There is a single hole injector mounted, which is in pre-chamber as shown in fig. 2-a. In addition, details of the computational mesh used in DI are given in Fig. 2-b. The computation used a 90 degree sector mesh (the diesel injector has four Nozzle holes) with 25 nodes in the radial direction, 20 nodes in the azimuthal direction and 5 nodes in the squish region (the region between the top of the piston and the cylinder head)at top dead center. The ground of the bowl has been meshed with two continuous layers for a proper calculation of the heat transfer through the piston wall. The final mesh consists of a hexahedral dominated mesh. Number of cells in the mesh was about 64,000 and 36,000 at BDC and TDC, respectively. The present resolution is found to give adequately at DI engine. Initial pressure in the combustion chamber is set to 86 kPa and initial temperature is calculated to be 384K, and swirl ratio is assumed to be on quiescent condition. Boundary temperatures for head, piston and cylinder are 550K, 590K and 450K, respectively. Present work is studied at full load mode and the engine speed is 730 rpm. All boundary temperatures were assumed to be constant throughout the simulation, but allowed to vary with the combustion chamber surface regions.

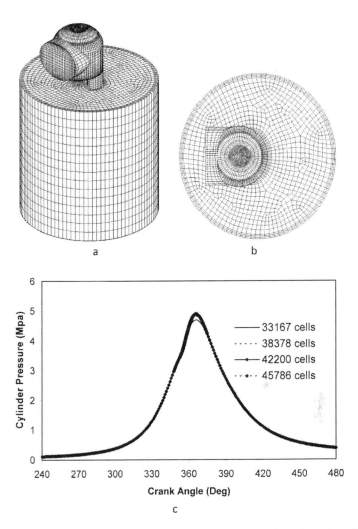

Figure 1. a. Mesh of the Lister 8.1 indirect injection diesel engine; b. Top view of the mesh; c. Grid dependency based on the in-cylinder pressure

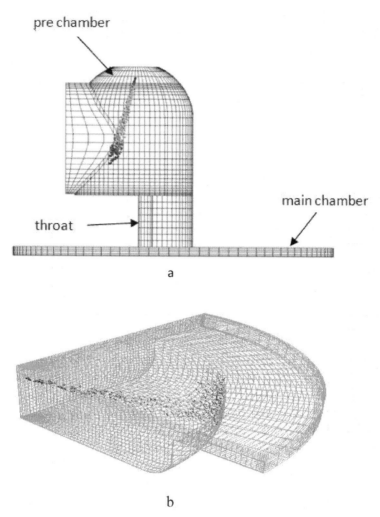

Figure 2. a. Spray and injector coordinate at pre-chamber; b Computational mesh with diesel spray drops at 350°CA, single injection case for DI engine.

3. MODEL FORMULATION

The numerical model is carried out for Lister 8.1 indirect injection diesel engine and OM355 DI engine with the specification given on tables 1 and 2, repectivily. The governing equations for unsteady, compressible, turbulent reacting multi-component gas mixtures flow and thermal fields are solved from IVC to EVO by the commercial AVL-FIRE CFD code [14]. The turbulent flow within the combustion chamber is simulated using the RNG k-ε turbulence model, modified for variable-density engine flows [15].

The standard WAVE model, described in [16], is used for the primary and secondary atomization modeling of the resulting droplets. At this model, the growth of an initial perturbation on a liquid surface is linked to its wavelength and other physical and dynamical parameters of the injected fuel at the flow domain. Drop parcels are injected with characteristic size equal to the Nozzle exit diameter (blob injection).

TABLE 1. Specifications of Lister 8.1 IDI diesel engine

Cycle Type	Four Stroke
Number of Cylinders	1
Injection Type	IDI
Cylinder Bore	114.1 mm
Stroke	139.7 mm
L/R	4
Displacement Volume	1.43 lit.
Compression Ratio	17.5 : 1
$V_{pre-chamber}/V_{TDC}$	0.7
Full Load Injected Mass	$6.4336e-5$ kg per Cycle
Power at 850 rpm	5.9 kW
Power at 650 rpm	4.4 kW
Initial Injection Pressure	90 bar
Nozzle Diameter at Hole Center	0.003m
Number of Nuzzle Holes	1
Nozzle Outer diameter	0.0003m
Spray Cone Angle	10°
Valve Timing	IVO= 5° BTDC
	IVC= 15° ABDC
	EVO= 55° BBDC
	EVC= 15° ATDC

The Dukowicz model is applied for treating the heat up and evaporation of the droplets, which is described in [17]. This model assumes a uniform droplet temperature. In addition, the droplet temperature change rate is determined by the heat balance, which states that the heat convection from the gas to the droplet either heats up the droplet or supplies heat for vaporization.

A Stochastic dispersion model was employed to take the effect of interaction between the particles and the turbulent eddies into account by adding a fluctuating velocity to the mean gas velocity. This model assumes that the fluctuating velocity has a randomly Gaussian distribution [14].

3. Model formulation

The spray/wall interaction model used in this simulation was based on the spray/wall impingement model [18]. This model assumes that a droplet, which hits the wall was affected by rebound or reflection based on the Weber number.

Table 2. Engine Specifications of OM-355 Diesel

Piston shape	Cylindrical bore
No. of nozzles/injector	4
Nozzle opening pressure	195 (bar)
Cylinders	6, In-line-vertical
Bore * stroke	128 (mm) * 150 (mm)
Max. power	179 (kw) at 2200 (rpm)
Compression ratio	16.1:1
Max. torque	824 N m at 1400 (rpm)
Capacity	11.58 (lit)
IVC	61°CA after BDC
EVO	60°CA before BDC
Initial Injection Pressure	250(bar)

The Shell auto-ignition model was used for modeling of the auto ignition [19]. In this generic mechanism, six generic species for hydrocarbon fuel, oxidizer, total radical pool, branching agent, intermediate species and products were involved. In addition, the important stages of auto ignition such as initiation, propagation, branching and termination were presented by generalized reactions, described in [14, 19].

The Eddy Break-up model (EBU) based on the turbulent mixing is used for modeling of the combustion in the combustion chamber [14]. This model assumes that in premixed turbulent flames, the reactants (fuel and oxygen) are contained in the same eddies and are separated from eddies containing hot combustion products. The rate of dissipation of these eddies determines the rate of combustion. In other words, chemical reaction occurs fast and the combustion is mixing controlled. NOx formation is modeled by the Zeldovich mechanism and Soot formation is modeled by Kennedy, Hiroyasu and Magnussen mechanism [20].

The injection rate profiles are rectangular type and consists of nineteen injection schemes, i.e. single injection and eighteen split injection cases(as shown in table 4). To simulate the split injection, the original single injection profile is divided into two injection pulses without altering the injection profile and magnitude. Fig. 3 illustrates the schematic scheme of the single and split injection strategy.

For the single injection case, the start of injection is at 348°CA and injection termination is at 387°CA. For all split injection cases, the injection timing of the first injection pulse is fixed at 348°CA. Three different split injection schemes, in which 10-20-25 % of total fuel injected in the second pulse, has been considered. The delay dwell between injections is varied from 5°CA to 30°CA with the interval 5°CA.

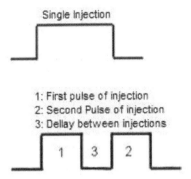

Figure 3. Schematic scheme of single and split injection strategy.

4. PERFORMANCE PARAMETERS

Indicated work per cycle is calculated from the cylinder pressure and piston displacement, as follows:

$$W = \int_{\theta_1}^{\theta_2} PdV \quad (1)$$

Where θ1, θ2 are the start and end of the valve-closed period, respectively (i.e. IVC= 15° ABDC and EVO= 55° BBDC). The indicated power per cylinder and indicated mean effective pressure are related to the indicated work per cycle by:

$$P(kW) = \frac{W(N.m)N(rpm)}{60000n} \quad (2)$$

$$IMEP = \frac{W}{V_d} \quad (3)$$

Where n=2 is the number of crank revolutions for each power stroke per cylinder, N is the engine speed in rpm and Vd is volume displacement of piston. The brake specific fuel consumption (BSFC) is defined as:

$$BSFC = \frac{m_f}{P_b}$$

(4)

In Equation (1), the work is only integrated as part of the compression and expansion strokes; the pumping work has not been taken into account. Therefore, the power and (ISFC) analyses can only be viewed as being qualitative rather than quantitative in this study.

5. RESULTS AND DISCUSSION IDI

Fig. 4 and Fig. 5 show the verification of computed and measured [21] mean in-cylinder pressure and heat release rate for the single injection case. They show that both computational and experimental data for cylinder pressure and heat release rate during the compression and expansion strokes are in good agreement.

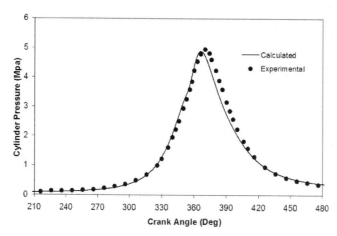

Figure 4. Comparison of calculated and measured [21] in-cylinder pressure, single injection case

The peak of cylinder Pressure is 4.88 Mpa, which occurs at 366°CA (4°CA after TDC). The start of heat release is at 351°CA for computed and measured results; in other words, the ignition delay dwell is 3°CA. It means that the ignition delay is quite close to the chemical ignition delay and that the physical ignition delay is very short, because of the rapid evaporation of the small droplets injected through the small injector gap at the start of injection. The heat release rate, which called "measured", is actually derived from the procured in-cylinder pressure data using a thermodynamic first law analysis as followed:

$$\frac{dq}{d\theta} = \frac{\gamma}{\gamma-1} p \frac{dV}{d\theta} + \frac{1}{\gamma-1} V \frac{dp}{d\theta} \tag{5}$$

Where p and V are in-cylinder pressure and volume versus the crank angle θ, and γ=1.33.The main difference of computed and measured HRR is due to applying single zone model to combustion process with assuming γ=1.33 and observed at premixed combustion. The peak of computed HRR is 59J/deg that occurs at 364°CA, compared to the peak of measured HRR that is 53J/deg at 369°CA. The main validation is based on pressure in cylinder.

As a whole, the premixed combustion occurs with a steep slope and it can be one of the major sources of NOx formation.

Table 3 shows the variation of performance parameters for the single injection case, compared with the experimental data [21]. In contrast with the experimental results, it can be seen that model can predict the performance parameters with good accuracy.

Fig. 6 indicates that the predicted total in-cylinder NOx emission for the single injection case, agrees well with the engine-out measurements [21]. Heywood [22] explains that the critical time for the formation of oxides of nitrogen in compression ignition engines is between the start of combustion and the occurrence of peak cylinder pressure when the burned gas temperatures are the highest. The trend of calculated NOx formation in the prechamber and main chambers agrees well with the Heywood's explanations. As temperature cools due to volume expansion and mixing of hot gases with cooler burned gas, the equilibrium reactions are quenched in the swirl chamber and main chamber.

As can be seen from the Fig. 7, the predicted total in-cylinder soot emission for the single injection case agrees well with the engine-out measurements [21].

5. Results and discussion IDI

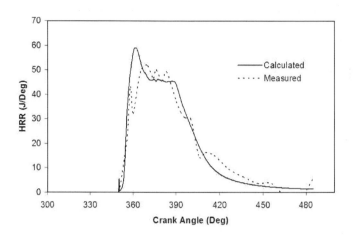

Figure 5. Comparison of calculated and measured [21] heat release rate, single injection case

Table 3. Comparison of calculated and measured [21] performance parameters, single injection case

parameters	Calculated	Experimental
Brake Power [kW]	4.53	4.65
BMEP [Bar]	5.33	5.47
Bsfc [g/kW.h]	310.72	302.7

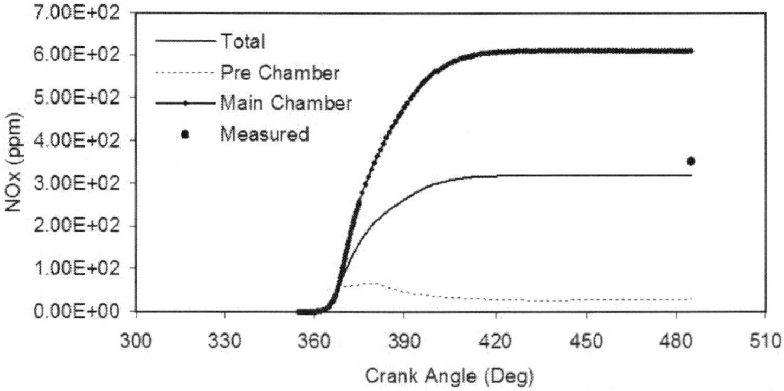

Figure 6. Comparison of calculated and measured [21] NOx emission, single injection case

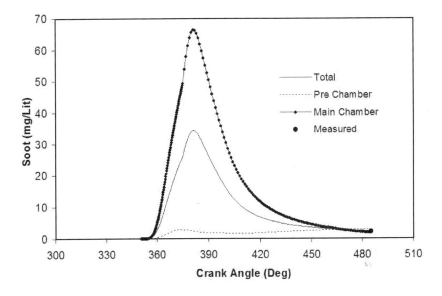

Figure 7. Comparison of calculated and measured [21] Soot emission, single injection case

Table 3 presents the exhaust NOx and soot emissions and performance parameters for the calculated single injection, and split injection cases. As can be seen, the lowest NOx and Soot emissions are related to the 75 %-15-25 % and 75 %-20-25 % cases respectively. In order to obtain the final optimum case, i.e. the case that involves the highest average of NOx and Soot reduction, a new dimensionless parameter is defined as:

The more of the total average emission reduction percentage results to the better optimum split injection case. Hence, it is concluded that the 75 %-20-25 % scheme with the average reduction percentage of 23.28 is the optimum injection case.

In addition, it can be deduced from the table 4 that in general, split injection sacrifices the brake power and Bsfc of the engine. This result is more apparent for the 80 %-20 % and 75 %-25 % cases. The reason is that with reduction of the first pulse of injection and increase of delay dwell between injections, premixed combustion as the main source of the power stroke is decreased. Hence, Bsfc is increased as well. As can be seen, the lowest brake power and highest Bsfc are related to the 75 %-30-25 % case.

Fig. 8 shows the NOx versus Soot emission for the single injection and split injection cases. As can be seen, the 75 %-20-25 % case is closer to origin, I,

e. zero emission. Hence, this confirms that the case of 75 %-20-25 % is the optimum case.

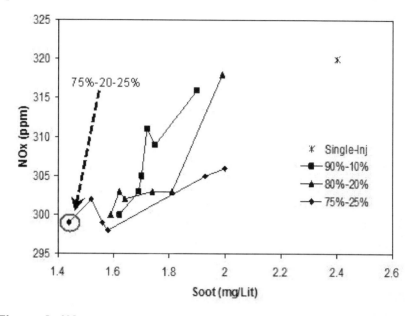

Figure 8. NOx versus soot emission for the single injection and split injection cases

It is of interest to notify that the optimum injection scheme for DI diesel engine at full load state is 75 %-25-25 % [11]. It means that the first and second injection pulses for the DI and IDI diesel engines are the same. The difference is related to the delay dwell between injections. I.e. the delay dwell for the optimum IDI split injection case is 5°CA lower than that of DI because of high turbulence intensity and fast combustion.

It is noticeable to compare the spray penetration, in-cylinder flow field, combustion and emission characteristics of the single injection and optimum injection cases to obtain valuable results.

The normalized injection profile versus crank angle for the single injection and 75 %-20-25 % cases is shown in the fig. 9. In this profile, actual injection rate values divided to maximum injection rate and this normalized injection rate profile is used by CFD code.

Fig. 10-a and Fig. 10-b represent respectively front and top views of the evolution of the spray penetration and velocity field at various crank angles in horizontal planes of the pre and main combustion chambers and planes across the connecting throat for the single injection and 75 %-20-25 %

cases. It can be seen that the maximum velocity in throat section are lower at all crank angles because of the large area this section than the other data in the literatures [23]. Generally, for all the cases, the maximum velocity of the flow field is observed at the tip of the spray, swirl chamber throat and some areas of the main chamber that is far from the cylinder wall and cylinder head.

Table 4. Exhaust emissions and performance parameters for the single injection and split injection cases

	NOx (ppm)	Soot (mg/lit)	NOx Reduction (%)	Soot Reduction (%)	Average reduction (%)	Pb (kW)	Bsfc (gr/kw.h)
Single Inj	320	2.4	0	0	0	4.53	310.72
90%-5-10%	316	1.9	1.25	20.83	11.04	4.51	312.1
90%-10-10%	309	1.75	3.43	27.08	15.25	4.4	319.9
90%-15-10%	300	1.62	6.25	32.5	19.37	4.32	325.8
90%-20-10%	311	1.72	2.81	28.33	15.57	4.37	322.1
90%-25-10%	303	1.69	5.31	29.58	17.44	4.33	325.08
90%-30-10%	305	1.7	4.68	29.16	16.92	4.35	323.58
80%-5-20%	318	1.99	0.6	17.08	8.84	4.36	322.84
80%-10-20%	303	1.81	5.31	24.58	14.94	4.31	326.58
80%-15-20%	303	1.74	5.31	27.5	16.4	4.2	335.14
80%-20-20%	302	1.64	5.62	31.66	18.64	4.14	340
80%-25-20%	303	1.62	5.31	32.5	18.9	4.13	340.82
80%-30-20%	300	1.59	6.25	33.75	20	4.12	341.65
75%-5-25%	306	2	4.37	20	12.18	4.29	328.11
75%-10-25%	305	1.93	4.68	19.58	12.13	4.18	336.74
75%-15-25%	298	1.58	6.87	34.16	20.51	4.12	341.65
75%-20-25%	299	1.44	6.56	40	23.28	4.04	348.41
75%-25-25%	302	1.52	5.62	36.66	21.14	4	351.9
75%-30-25%	299	1.56	6.56	35	20.78	3.97	354.55

5. Results and discussion IDI

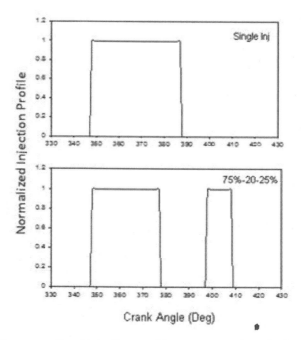

Figure 9. The normalized injection profile versus crank angle for the single injection and 75 %-20-25 % cases

As can be seen, at various crank angles, the main difference of the in-cylinder flow field between the single injection and split injection cases is due to the fuel injection scheme. In other words, the amount of the fuel spray and the crank position in which the spray is injected. The aerodynamic forces decelerate the droplets for the both cases. The drops at the spray tip experience the strongest drag force and are much more decelerated than droplets that follow in their wake.

At 370°CA, air entrainment into the fuel spray can be observed for the both cases. Hence, Droplet velocities are maximal at the spray axis and decrease in the radial direction due to interaction with the entrained gas.

Although the amount of the fuel spray for the single injection case is higher than the 75 %-20-25 % case at 370°CA, the flow field difference is not observed obviously. The more quantity of the fuel spray for the single injection case causes the maximum velocity of the single injection case to be higher by about 1m/s than the 75 %-20-25 % case.

At 380°CA, the fuel spray is cut off for the 75 %-20-25 % case. Since the entrained gas into to the spray region is reduced, the flow moves more

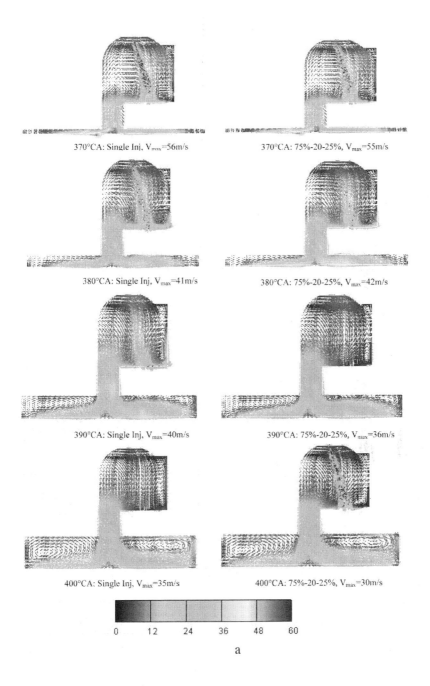

370°CA: Single Inj, V_{max}=56m/s	370°CA: 75%-20-25%, V_{max}=55m/s
380°CA: Single Inj, V_{max}=41m/s	380°CA: 75%-20-25%, V_{max}=42m/s
390°CA: Single Inj, V_{max}=40m/s	390°CA: 75%-20-25%, V_{max}=36m/s
400°CA: Single Inj, V_{max}=35m/s	400°CA: 75%-20-25%, V_{max}=30m/s

a

5. Results and discussion IDI

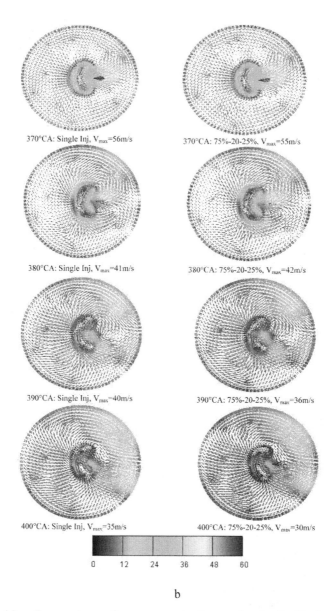

b

Figure 10 a. Comparison of spray penetration and velocity filed at various crank angles for the single injection and 75 %-20-25 % cases, front view; b. Comparison of spray penetration and velocity filed at various crank angles for the single injection and 75 % -20-25 % cases, top view.

freely from the swirl chamber to the main chamber. Hence, as can be seen from the front view, the maximum flow field velocity for the 75 %-20-25 % case is higher compared to the single injection case.

At 390°CA, from the top view, the distribution of the flow field in the whole of the main chamber is visible. In addition, some local swirls can be seen in the main chamber that are close to the swirl chamber. From the front view, it is observed that the gas coming from the pre-chamber reaches the opposite sides of cylinder wall. This leads to the formation of the two eddies occupying each one-half of the main chamber and staying centered with respect to the two half of the bowl.

The start of the second injection pulse for the 75 %-20-25 % case is observed At 400°CA. since the swirl intensity in the swirl chamber is reduced at this crank position, the interaction between the flow field and spray is decreased to somehow and flow in the pre-chamber is not strongly influenced by the fuel spray. Hence, the maximum velocity of the flow field for the 75 %-20-25 % case remains lower compared to the single injection case.

Fig. 11 indicates the history of heat release rate, cylinder pressure, temperature, O_2 mass fraction, NOx and soot emissions for the single injection and 75 %-20-25 % cases.

Fig. 11-a shows that the amount of heat release rate for the both cases is to somehow equal until 380°CA. It is due to the fact that the first injection pulse for the 75 %-20-25 % case, lasts to 377°CA. Hence, the premixed combustion for the both cases does not differ visibly. For the 75 %-20-25 % case, The second peak of heat release rate occurs at 410°CA and indicates that a rapid diffusion burn is realized at the late combustion stage and it affects the in-cylinder pressure, temperature and soot oxidation.

Fig. 11-b shows that the second injection pulse of the 75 %-20-25 % does not cause to the second peak of the cylinder pressure. It is because the rate of decrease of cylinder pressure due to the expansion stroke is almost equal to the rate of increase of cylinder pressure due to the diffusion combustion.

Fig. 11-c compares temperature trend for 75 %-20-25 % and single injection cases. As can be seen, for the 75 %-20-25 % case, the peak of the temperature advances about 11°CA compared to the peak of the temperature of the single injection case. In addition, the temperature reduction reaches to 90°K.

Fig. 11-d presents the in cylinder O_2 mass fraction. It can be seen that after 385°CA, the O2 concentration for the 75 %-20-25 % case is higher than that for the single injection case. In other words, oxygen availability of 75 %-20-25 % case is better.

5. Results and discussion IDI

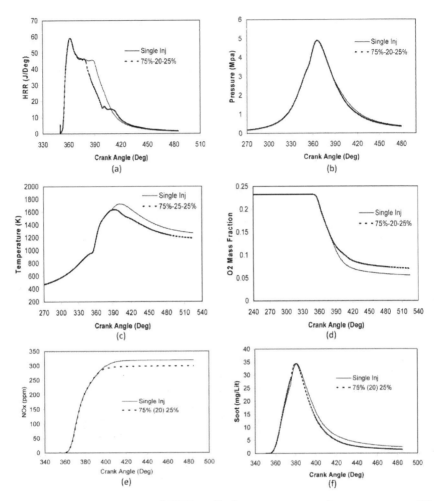

Figure 11. Comparison of HRR, cylinder pressure and temperature, O2 mass fraction, NOx and soot histories for the single injection and 75 %-20-25 % cases

As Fig. 11-e and Fig. 11-f indicate, the reduction of the peak of the cylinder temperature for the 75 %-20-25 % case, causes to the lower NOx formation. In addition, soot oxidation for the 75 %-20-25 % case is higher compared to that for the single injection case. It is because for the 75 %-20-25 % case, the more availability of oxygen in the expansion stroke compensates the decrease of temperature due to the split injection. Hence, soot oxidation is increased.

Fig. 12-a and Fig. 12-b represent respectively front and top views of the contours of equivalence ratio, temperature, NOx and soot emissions at various crank angles in horizontal planes of the main combustion chamber and planes across the connecting throat for the single injection and 75 %-20-25 % cases.

At 400°CA, for the both cases, the majority of NOx is located in the two half of the main chamber, whilst soot is concentrated in the swirl chamber, throat section and main chamber. For the both cases, a local soot-NOx trade-off is evident in the swirl chamber, as the NOx and soot formation occur on opposite sides of the high temperature region. Equivalence ratio contour of the 75 %-20-25 % case confirms that injection termination and resumption, causes leaner combustion zones. Hence, the more reduction of soot for the 75 %-20-25 % is visible compared to the single injection case.

NOx and soot formation depend strongly on equivalence ratio and temperature. It is of interest to notice that the area which the equivalence ratio is close to 1 and the temperature is higher than 2000 K is the NOx formation area. In addition, the area which the equivalence ratio is higher than 3 and the temperature is approximately between 1600 K and 2000 K is the Soot formation area. The area from 1500 K and equivalence ratio near to 1 is defined as soot oxidation area [24, 25].

At 410°CA, due to the second injection pulse, increase of the equivalence ratio is observed in the swirl chamber of the 75 %-20-25 % case. Hence, the enhancement of the soot concentration close to the wall spray impingement is observed. For the other areas, soot oxidation for the 75 %-20-25 % case is higher compared to the single injection case. In addition, NOx reduction tendency is visible in the right side of the main chamber of the 75 %-20-25 % case compared to the single injection case. It is because more temperature reduction occurs for the 75 %-20-25 % case due to the lower premixed combustion. The diffusion combustion of the second injection pulse does not affect NOx formation significantly.

The final form of the distribution of equivalence ratio, temperature, NOx and soot contours can be seen at 420°CA. The reduction of soot and NOx for the 75 %-20-25 % case is clear compared to the single injection case. This trend is preserved for the both cases until EVO.

5. Results and discussion IDI

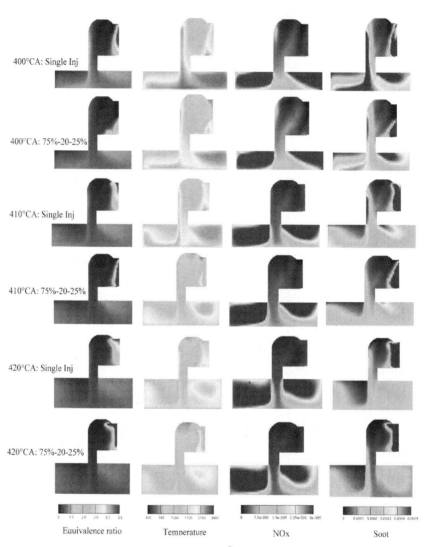

a

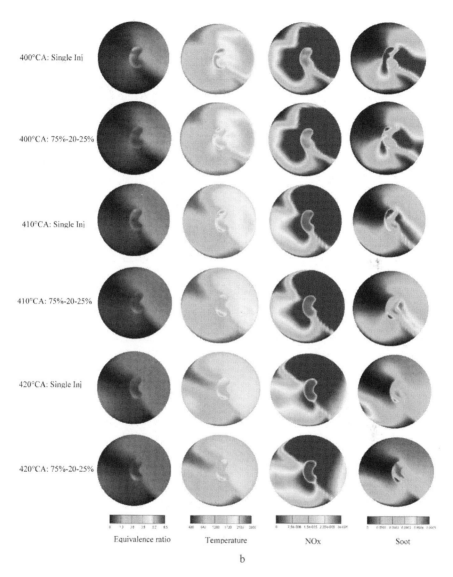

Figure 12 a. Contours of equivalence ratio, temperature, NOx and Soot at different crank angles, front view b. Contours of equivalence ratio, Temperature, NOx and Soot at different crank angles, top view

6. RESULTS AND DISCUSSION OF DI DIESEL ENGINE

Fig. 13 shows the cylinder pressure and the rate of heat release for the single injection case. As can be seen from HRR curve, the peak of the heat release rate occurs at 358ºCA (2ºCA before TDC). The premixed combustion occurs with a steep slope and it can be one of the major sources of NOx formation. The good agreement of predicted in-cylinder pressure with the experimental data [27] can be observed.

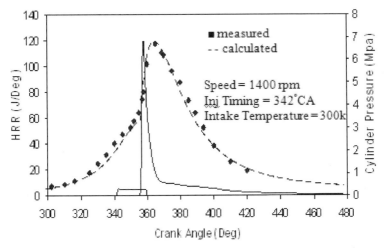

Figure 13. HRR and Comparison of calculated and measured [27] in-cylinder pressure, single njection case.

Fig. 14 and Fig. 15 imply that the predicted total in-cylinder NOx and soot emissions for the single injection case, agree well with the engine-out measurements [27].

Fig. 16 shows the trade-off between NOx and soot emissions at EVO when the injection timing is varied. As indicated, the general trends of reduction in NOx and increase in soot when injection timing is retarded can be observed and it is independent on injection strategy. The reason is that it causes the time residence and ignition delay to be shorter, resulting in a less intense premixed burn and soot formation increases; in addition, the less temperature in different parts of combustion chamber keeps the soot oxidation low but decreases the formation of thermal NOx.

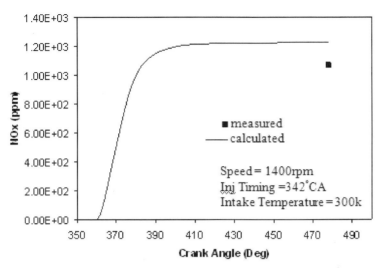

Figure 14. Comparison of calculated and measured [27] NOx emission, single injection case.

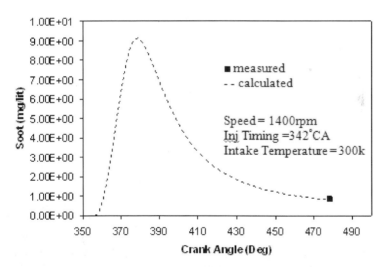

Figure 15. Comparison of calculated and measured [27] soot emission, single injection case.

6. Results and discussion of DI diesel engine

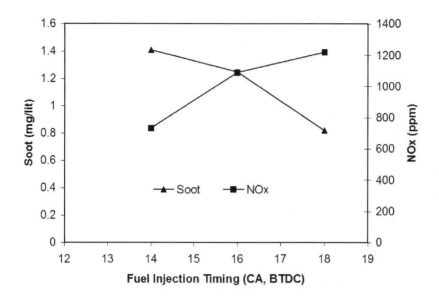

Figure 16. The effect of injection timing on NOx and Soot trade-off, single injection case.

To simulate the split injection, the original single injection profiles are split into two injection pulses without altering the injection profile and magnitude. In order to obtain the optimum dwell time between the injections, three different schemes including 10 %, 20 % and 25 % of the total fuel injected in the second pulse are considered.

Fig. 17 shows the effect of delay dwell between injection pulses on soot and NOx emissions for the three split injection cases. For all the cases, the injection timing of the first injection pulse is fixed at 342ºCA.

The variation trend of curves in Fig. 17 is very similar to the numerical results obtained by Li et al [28]. As can be seen, the optimum delay dwell between the injection pulses for reducing soot with low NOx emissions is about 25ºCA. The evidences of Li, J. et al. [28], which has used a phenomenological combustion model, support this conclusion. Table 5 compares Exhaust NOx and soot emissions for the single injection and optimum split injection cases for the optimum delay dwell. As shown, for the 75 % (25) 25 % case, NOx and soot emissions are lower than the other cases. It is due to the fact that the premixed combustion which is the main source of the NOx formation is relatively low. The more quantity of the second injection pulse into the lean and hot combustion zones, causing the newly injected fuel to burn rapidly and efficiently at high temperatures resulting in high soot oxidation rates. In addition, the heat released by the second injection pulse is not sufficient to increase the NOx emissions. Fig. 18 and Fig. 19 confirm the explanations.

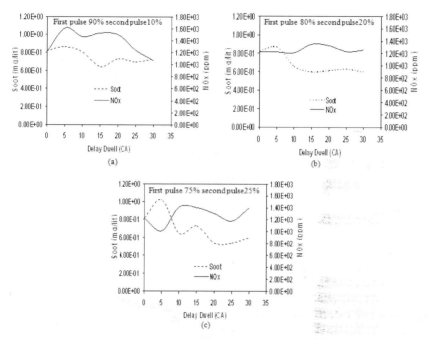

Figure 17. The variation of soot and NOx at different delay dwells, split injection cases.

Table 5. Comparison of NOx and soot emissions among the single injection and optimum split injection cases for the optimum delay dwell (25ºCA)

Case	NOx (ppm)	Soot (mg/lit)
Single inj	1220	0.82
Split inj- 90% (25) 10%	1250	0.697
Split inj- 80% (25) 20%	1230	0.614
Split inj- 75% (25) 25%	1180	0.541

Fig. 18 shows the cylinder pressure and heat release rates for the three split injection cases in the optimum delay dwell i.e. 25ºCA. As shown, split injection reduces the amount of premixed burn compared to the single injection case in Fig. 13. The second peak which appears in heat release rate curves of split injection cases indicates that a rapid diffusion burn is realized at the late combustion stage and it affects the in-cylinder pressure and temperature. The calculated results of the cylinder pressure and HRR for the optimum split injection cases show very good similarity with the results of the reference[28].

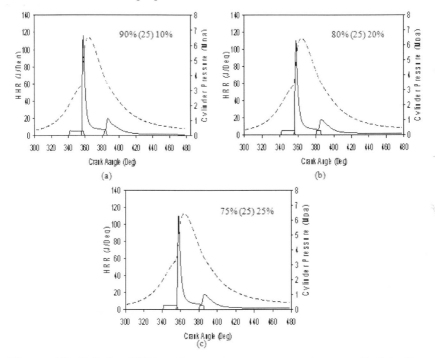

Figure 18. Cylinder HRR and pressure curves, optimum split injection cases.

Fig. 19 indicates the cylinder temperature for the single injection and optimum split injection cases. For the split injection cases, the two peaks due to the first and second injection pulses in contrast with the one peak in the single injection case can be observed. As shown, for the 75 % (25) 25 % case, the first and second peaks which are related to the premixed combustion and NOx formation are lower than the other cases. In addition, after the second peak, the cylinder temperature tends to increase more in comparison with the other cases and causes more soot oxidation. Hence, for the 75 % (25) 25 % case, NOx and soot emissions are lower than the other optimum cases.

Figure.20. shows the isothermal contour plots at different crank angle degrees for the 75 % (25) 25 % case at a cross-section just above the piston bowl. The rapid increase in temperature due to the stoichiometric combustion can be observed at 355ºCA. At 360ºCA, the injection termination can be observed which the in-cylinder temperature tends to be maximum. At 370ºCA, the fuel injection has been cut off and the cylinder temperature tends to become lower. As described, injection termination and resumption prevents not only fuel rich combustion zones but also causes more complete combustion due to better air utilization. The resumption of the injection can be observed at 380ºCA which causes the diffusion combustion and increases the temperature in the cylinder. At 390ºCA, the increase of cylinder temperature creates bony shape contours and soot oxidation increases.

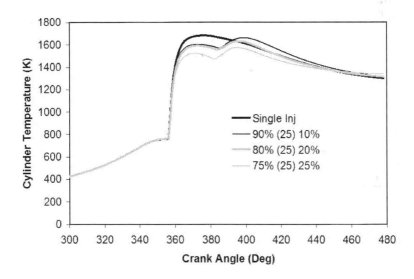

Figure 19. Comparison of cylinder temperature among the single injection and optimum split injection cases.

6. Results and discussion of DI diesel engine

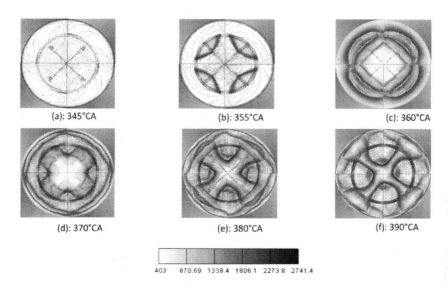

Figure 20. Isothermal contour plots of 75 % (25) 25 % case at different crank angle degrees.

Fig. 21 compares the contour plots of NOx, temperature, equivalence ratio and soot for the single injection and 75 % (25) 25 % cases in a plane through the center of the spray at 380ºCA. As explained above, 380ºCA corresponds to a time when the second injection pulse has just started for the 75 % (25) 25 % case. For the two described cases, it can be seen that the area which the equivalence ratio is close to 1 and the temperature is higher than 2000 K is the NOx formation area. In addition, the area which the equivalence ratio is higher than 3 and the temperature is approximately between 1600 K and 2000 K is the soot formation area. A local soot-NOx trade-off is evident in these contour plots, as the NOx formation and soot formation occur on opposite sides of the high temperature region. It can be seen that for the 75 % (25) 25 % case, NOx and soot mass fractions are lower in comparison with the single injection case. Because of the optimum delay dwell, the second injection pulse, maintains the low NOx and soot emissions until EVO.

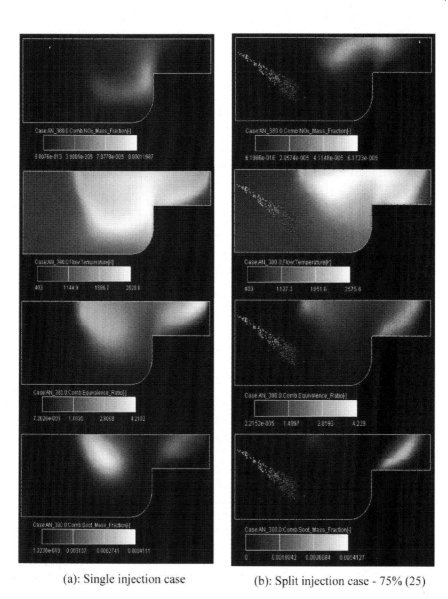

(a): Single injection case   (b): Split injection case - 75% (25)

Figure 21. Contour plots of NOx, temperature, equivalence ratio, and soot with fuel droplets at 380°CA for the single injection and 75 % (25) 25 % cases.

6. CONCLUSION

At the present work, the effect of the split injection on combustion and pollution of DI and IDI diesel engines was studied at full load state by the CFD code. The target was to obtain the optimum split injection cases for these engine in which the total exhaust NOx and soot concentrations are the lowest.

Three different split injection schemes, in which 10, 20 and 25 % of total fuel injected in the second pulse, was considered. The delay dwell between injections pulses is varied from 5°CA to 30°CA with the interval 5°CA. The results for IDI are as followed:

1. The calculated combustion and performance parameters, exhaust NOx and soot emissions for the single injection case showed a good agreement with the corresponding experimental data.

2. The lowest NOx and Soot emissions are related to the 75 %-15-25 % and 75 %-20-25 % cases respectively. Finally, optimum case was 75 %-20-25 % regarding the highest average of NOx and soot reduction.

3. The lowest brake power and highest BSFC quantity is due to the 75 %-30-25 % case.

4. Because in the literature, the optimum split injection scheme for DI diesel engines at full state was defined as 75 %-25-25 %, it was concluded that the difference of the optimum split injection scheme for DI and IDI diesel engines was related to the delay dwell between the injections.

5. The main difference of the in-cylinder flow field between the single injection and split injection cases is due to the fuel injection scheme.

6. For the 75 %-20-25 % case, the more availability of oxygen in the expansion stroke compensates the decrease of temperature due to the split injection. Hence, soot oxidation is increased.

7. The final form of the NOx and soot at 420°CA which preserve it's trend until EVO, confirms the more reduction of NOx and soot for the 75 %-20-25 % case in comparison with the single injection case.

In addition, the results for DI engins are as follow:

1. A good agreement of predicted in-cylinder pressure and exhaust NOx and soot emissions with the experimental data can be observed.

Diesel Engine - Combustion, Emissions and Condition Monitoring

2. Advancing or retarding the injection timing can not decrease the soot and NOx trade-off by itself. Hence split injection is needed.

3. The optimum delay dwell between the injection pulses for reducing soot with low NOx emissions is about 25ºCA. The results of phenomenological combustion models in the literature support this conclusion.

4. The calculated results of the cylinder pressure and heat release rate for the optimum split injection cases show very good similarity with the numerical results obtained by phenomenological combustion models.

5. For the 75 % (25) 25 % case, NOx and soot emissions are lower than the other cases. It is due to the fact that the premixed combustion which is the main source of the NOx formation is relatively low. The more quantity of the second injection into the lean and hot combustion zones, causing high soot oxidation rates. In addition, the heat released by the second injection pulse is not sufficient to increase the NOx emissions.

6. Contour plots of NOx, temperature, equivalence ratio and soot for the single injection and 75 % (25) 25 % cases at 380ºCA show that the area which the equivalence ratio is close to 1 and the temperature is higher than 2000 K is the NOx formation area. In addition, the area which the equivalence ratio is higher than 3 and the temperature is approximately between 1600 K and 2000 K is the Soot formation area.

REFERENCES

1. M Gomaa, A. J Alimin, K. A Kamarudin, Trade-off between NOx, soot and EGR rates for an IDI diesel engine fuelled with JB5.World Academy of ScienceEngineering and Technology 201062449450

2. G. M Bianchi, P Peloni, F. E Corcione, F Lupino, Numerical analysis of passenger car HSDI diesel engines with the 2nd generation of common rail injection systems: The effect of multiple injections on emissions. SAE Paper 2001-01

3. Seshasai SrinivasanFranz X. Tanner, Jan Macek and Milos Polacek. Computational Optimization of Split Injections and EGR in a Diesel Engine Using an Adaptive Gradient-Based Algorithm. SAE paper, 2006

REFERENCES

4. Shayler PjNg HK. Simulation studies of the effect of fuel injection pattern on NOx and soot formation in diesel engines. SAE Paper 2004-01

5. C. A Chryssakis, D. N Assanis, S Kook, C Bae, Effect of multiple injections on fuel-air mixing and soot formation in diesel combustion using direct flame visualization and CFD techniques. Spring Technical Conference, ASME ICES2005-101620051016

6. GA Lechner, , TJ Jacobs, , CA Chryssakis, , DN Assanis, , RM Siewert, . Evaluation of a narrow spray cone angle, advanced injection SAE Paper trategy to achieve partially premixed compression ignition combustion in a diesel engine. SAE Paper 2005-01-0167; 2005.

7. R Ehleskog,. Experimental and numerical investigation of 2006plit injections at low load in an HDDI diesel engine equipped with a piezo injector. SAE Paper 2006-01-3433; 2006.

8. YD Sun, R Reitz, . Modeling diesel engine NOx and soot reduction with optimized two-stage combustion. SAE Paper 2006-01-0027; 2006.

9. K Verbiezen, et alDiesel combustion: in-cylinder NO concentrations in relation to injection timing.Combustion and Flame, 2007151333346

10. N Abdullah, A Tsolakis, P Rounce, M Wyszinsky, H Xu, R Mamat, Effect of injection pressure with split injection in a 6diesel engineSAE Paper 2009-242009

11. S Jafarmadar, A Zehni, Multi-dimensional modeling of the effects of split injection scheme on combustion and emissions of direct-injection diesel engines at full load stateIJE, 2009224

12. K Showry, A Raju, Multi-dimensional modeling and simulation of diesel engine combustion using multi-pulse injections by CFD. International Journal of Dynamics of Fluids 20100973-178462237248

13. K Iwazaki, K Amagai, M Arai, Improvement of fuel economy of an indirect (IDI) diesel engine with two-stage injection. Energy Conversion and Management, 200530447459

14. AVL FIRE user manual V2006

15. Z Han, R D Reitz, Turbulence modeling of internal combustion engines using RNG K-ε models. Combustion Science and Technology 1995267295

16. A. B Liu, R. D Reitz, Modeling the effects of drop drag and break-up on fuel sprays. SAE Paper 9300721993

17. J. K Dukowicz, Quasi-steady droplet change in the presence of convection. Informal report Los Alamos Scientific Laboratory. LA7997MS.

18. J. D Naber, R. D Reitz, Modeling engine spray/wall impingementSAE Paper 8801071988

19. M Halstead, L Kirsch, C Quinn, The Auto ignition of hydrocarbon fueled at high temperatures and pressures- fitting of a mathematical model. Combustion Flame 1977304560

20. M. A Patterson, S. C Kong, G. J Hampson, R. D Reitz, Modeling the effects of fuel injection characteristics on diesel engine soot and NOx emissionsSAE Paper 9405231994

21. Khoushbakhti Saray RMohammahi Kusha A, Pirozpanah V. A new strategy for reduction of emissions and enhancement of performance characteristics of dual fuel engine at part loads. Int. J. Engineering 20102387104

22. J. B Heywood, Internal combustion engine fundamental. New York: McGraw Hill Book Company; 1988568586

23. S Sbastian, S Wolfgang, Combustion in a swirl chamber diesel engine simulation by computation of fluid dynamics. SAE Paper 9502801995

24. H Yoshihiro, N Kiyomi, I Minaji, Combustion improvement for reducing exhaust emissions in IDI diesel engineSAE Paper 98050319 EOF31 EOF1998

25. S Hajireza, G Regner, A Christie, M Egert, H Mittermaier, Application of CFD modeling in combustion bowl assessment of diesel engines using DoE methodology. SAE Paper 2006-012006

26. B Carsten, Mixture formation in internal combustion enginesBerlin: Springer publications; 2006226231

27. V Pirouzpanah, and B. O Kashani, Prediction of major pollutants emission in direct-injection dual-fuel diesel and natural gas enginesSAE Paper, 1999-011999

28. J Li, J Chae, S Lee, and J. S Jeong, Modeling the effects of split injection scheme on soot and NOx emissions of direct injection diesel engines by a phenomenological combustion model", SAE Paper, 9620621996

CHAPTER 2

Analytical Methodologies for the Control of Particle-Phase Polycyclic Aromatic Compounds from Diesel Engine Exhaust

F. Portet-Koltalo and N. Machour

[1] UMR CNRS 6014 COBRA, Université de Rouen, Evreux, France

1. INTRODUCTION

Diesel emissions contain complex mixtures of chemical constituents that are known to be (or possibly to be) human carcinogens, or that have adverse health effects [1]. Among these substances (formaldehyde, benzene, acrolein, dioxins, etc.), polycyclic aromatic hydrocarbons (PAHs) and their nitrated derivatives (nitro-PAHs) have been implicated as major contributors to the toxicity of diesel exhausts [2, 3]. However, these substances are currently non-regulated pollutants in engine exhaust. PAHs are a group of organic compounds consisting of two or more fused aromatic rings, which are produced as a result of the incomplete combustion of fossil fuels. These compounds are mainly derived from anthropogenic sources, particularly from mobile sources of emissions in urban areas [4]. The combustion of fuel in diesel engines results in the formation of a mixture of gaseous compounds (CO, CO_2, NO, NO_2, SO_2, etc.) and solid particles (carbonaceous matter, sulphates, trace metals, etc.). PAHs and their derivatives can be found in the gaseous or particulate phases of diesel exhaust fumes depending on their vapour pressure, but the distribution of particles between the two phases is also dependent on the total amount of particles, on the physical characteristics of the particles (particularly their specific surface area), and on the temperature.

The majority of particulate matter in diesel exhaust is composed of fine particles that are primarily formed from the condensation of organic matter on an elemental carbon core. These particles are generally called soot particles. The soluble organic fraction (SOF) is defined as the fraction that can be extracted from soot using an organic solvent; particulate PAHs and their derivatives are found in this fraction. The SOF primarily originates from unburned fuel and engine lubrication oil [5]. There is a wide range of SOF in the diesel particulate matter; it can range from less than 10 % to more than 90 % by soot mass. The fraction of SOF in the particulate matter depends on the type of diesel vehicles (light or heavy-duty diesel vehicles, mopeds, etc.), on the concentrations of sulfur and aromatic compounds in the diesel fuel, on the test engine cycles and also, on the collecting procedure of diesel exhaust, which is less well described. In general, the SOF values are highest for light engine loads when exhaust temperatures are low [6].

Two methods are generally used for measuring vehicle emissions: dynamometer studies where tailpipe fumes are measured from one vehicle, and roadside or tunnel studies where mean traffic emissions values (daily, by vehicles, etc.) are obtained [7]. In this book chapter, only analytical methodologies developed for diesel emissions studies using dynamometer tests are described. Even if the exhaust particles produced in the tailpipe are distinct from those sampled in the ambient atmosphere, it is shown in this chapter that the trapping media are very similar. Conventional and less conventional sampling trains for collecting the particles are described. The classical (Soxhlet and ultra-sound extractions) and more recent methodologies (micro-wave assisted extractions, pressurised solvent extractions, supercritical fluid extractions, and solid phase extractions) are detailed; they are used for treating the collected samples from gaseous and particulate phases to obtain the fraction containing PAHs and their derivatives. Because the individual products present different health risks, information about total PAH emissions is less important than information about the composition of such emissions. Therefore, techniques that are developed for identifying and quantifying the individual PAHs and their nitrated derivatives found in the two phases are described.

Finally, the efficiency of the sample pretreatment step (before quantitation) is strongly dependent on the SOF fraction of diesel soot; therefore, it is shown in this chapter that when the optimisation of this important analytical step is neglected, it can lead to important analytical bias.

2. PROPERTIES, SOURCES AND TOXICITY OF PAHS AND NITRO-PAHS

2.1. Physical and chemical properties of pahs and nitro-pahs

The name "polycyclic aromatic hydrocarbons" (PAH) commonly refers to a large class of organic compounds containing two or more fused aromatic rings; however, in a broad sense, non-fused ring systems should also be included. In particular, the term "PAH" refers to compounds containing only carbon and hydrogen atoms (*i.e.,* unsubstituted parent PAH and their alkyl-substituted derivatives), whereas the more general term "polycyclic aromatic compounds" also includes the functional derivatives (*e.g.,* nitro- and hydroxy-PAHs) and the heterocyclic analogues, which contain one or more hetero atoms in the aromatic structure (aza-, oxa-, and thia-arenes).

The physical and chemical properties of PAHs are dependent on the number of aromatic rings and on the molecular mass (Table 1). The smallest member of the PAH family is naphthalene, a two-ring compound, which is mainly found in the vapour phase in the atmosphere, because of its higher vapour pressure and Henry's law constant K_H (ratio between the partial pressure of a gas above a solution and the amount of gas solubilized in solution). Three to five ring-PAHs can be found in both the vapour and particulate phases in air. PAHs consisting of five or more rings tend to be solids adsorbed onto other particulate matter in the atmosphere. For instance, the resistance of PAHs to oxidation, reduction, and vaporisation increases with increasing molecular weight, whereas the aqueous solubility of these compounds decreases (their *n*-octanol/water partition coefficient log K_{ow} is higher). As a result, PAHs differ in their behaviour, their distribution in the environment, and their effects on biological systems.

2.2. Toxicity of Pahs and Nitropahs

Over 100 chemical compounds formed during the incomplete combustion of organic matter are classified as PAHs. These ubiquitous environmental pollutants are human carcinogens and mutagens; therefore, they are toxic to all living organisms. There are various pathways for human exposure to PAHs. For the general population, the major routes of exposure are from food and inhaled air; however, in smokers, the exposure from smoking and food may be of a similar magnitude. Extensive reviews and guidelines concerning PAHs contamination in food and air have been performed, and they have questioned the correct marker capable of following the risk assessment for the population [10].

2. Properties, sources and toxicity of PAHs and nitro-PAHs

Table 1. Physical and chemical properties of some PAHs and nitro-PAHs [8, 9].

Compounds/ Abbreviations	Chemical structure	Molecular mass (g mol⁻¹)	Melting/ Boiling points (°C)	Vapour Pressure (Pa at 25°C)	Log K_{ow}	Solubility in water at 25°C (mg L⁻¹)	K_H at 25°C (Pa.m³ mol⁻¹)
Naphthalene* NAPH		128.18	81 / 218	10.4	3.4	31.7	49
Acenaphtylene* ACY		152.18	92 / 270	0.89	4.07	-	114
Acenaphthene* ACE		154.21	95 / 279	0.29	3.92	3.9	15
Fluorene* FLUO		166.22	116 / 295	0.08	4.18	1.68	9.81
Anthracene* ANT		178.24	216.4 / 342	8.0 10⁻⁴	4.5	0.073	73
Phenanthene* PHEN		178.24	100.5/ 340	0.016	4.52	1.29	4.29
Fluoranthene* FLT		202.26	108.8/ 375	0.00123	5.20	0.26	1.96
Pyrene* PYR		202.26	150.4 / 393	0.0006	5.18	0.135	1.1
Chrysene* CHRYS		228.29	253.8 / 436	-	5.86	0.00179	0.53
Benz[a] Anthracene* B(a)ANT		228.29	160.7 / 400	2.8 10⁻⁵	5.61	0.014	1.22
Benzo[b] Fluoranthene* B(b)FLT		252.32	168.3 / 481	-	5.78	0.0015	0.051
Benzo[k] Fluoranthene* B(k)FLT		252.32	215.7 / 480	-	6.11	0.0008	0.044
Benzo[a]pyrene* B(a)PYR		252.32	178.1 / 496	7.3 10⁻⁷	6.50	0.004	0.034 (20°C)
Benzo[e]pyrene B(e)PYR		252.32	178.7 / 493	7.4 10⁻⁷	6.44	0.005	-

Compound	Structure	MW	mp/bp	Vapor pressure	log Kow	Col 7	Col 8
Perylene PER		252.32	277.5 / 503	-	5.3	0.0004	-
Benzo[g,h,i]Perylene* B(ghi)PER		276.34	278.3 / 545	$1.4 \cdot 10^{-8}$	7.10	0.00026	0.027 (20°C)
Indeno[1,2,3-cd]Pyrene* InPYR		276.34	163.6 / -	-	-	0.00019	0.029 (20°C)
Dibenz[a,h]Anthracene* DB(ah)ANT		278.35	266.6 / -	-	6.75	0.00050	-
Coronene COR		300.36	439 / 525	$2 \cdot 10^{-10}$	5.4	0.00014	-
2-nitronaphthalene 2N-NAPH		173.17	74 / 304	$3.2 \cdot 10^{-2}$	2.78	26	610
2-nitrofluorene 2N-FLUO		211.22	154 / 326	$9.7 \cdot 10^{-5}$	3.37	0.216	95
Pyrene* PYR		202.26	150.4 / 393	0.0006	5.18	0.135	1.1
Chrysene* CHRYS		228.29	253.8 / 436	-	5.86	0.00179	0.53
Benz[a]Anthracene* B(a)ANT		228.29	160.7 / 400	$2.8 \cdot 10^{-5}$	5.61	0.014	1.22
Benzo[b]Fluoranthene* B(b)FLT		252.32	168.3 / 481	-	5.78	0.0015	0.051
Benzo[k]Fluoranthene* B(k)FLT		252.32	215.7 / 480	-	6.11	0.0008	0.044
Benzo[a]pyrene* B(a)PYR		252.32	178.1 / 496	$7.3 \cdot 10^{-7}$	6.50	0.004	0.034 (20°C)
Benzo[e]pyrene B(e)PYR		252.32	178.7 / 493	$7.4 \cdot 10^{-7}$	6.44	0.005	-

Compound	Structure	MW					
9-nitroanthracene 9N-ANT		223.23	146 / -	-	4.16	-	-
2-nitrophenanthrene 2N-PHEN		223.23	120 / -	-	4.23	-	-
2-nitrofluoranthene 2N-FLT		247.25	- / 420	$9.9 \cdot 10^{-7}$	-	0.019	13
1-nitropyrene 1N-PYR		247.25	152 / 472	$4.4 \cdot 10^{-6}$	5.29	0.017	64
2-nitropyrene 2N-PYR		247.25	199 / 472	$4.4 \cdot 10^{-6}$	-	0.021	64
7-nitrobenz[a]anthracene 7N-B(a)ANT		273.29	162 / -	-	5.34	-	-
6-nitrobenzo[a]pyrene 6N-B(a)PYR		297.31	260 / 567	-	6.13	-	12

* 16 priority PAHs defined by the US-EPA

There have been concerns regarding the carcinogenic effects of these products for at least two centuries, following the report of a high incidence of scrotal cancer in soot workers in London by Sir P. Pottwatched. The association between cancer and a specific chemical compound, Benz[a]pyrene, was established when this compound was isolated from chimney soot. Several subsequent studies proved that, in addition to Benz[a]pyrene, other PAHs were also carcinogenic [11]. The United States Environmental Protection Agency (U.S. EPA) targeted sixteen specific PAHs for measurements in environmental samples (see Table 1), and Benz[a]pyrene (indicator of PAHs species) was classified as a 2B pollutant, which means that it is a probable human carcinogen based on sufficient evidence from animal studies, but there is inadequate evidence from human studies [12]. According to the International Agency for Research on Cancer Classification (IARC), it is Group 2A compound, which means that it is likely carcinogenic to humans [13]. The World Health Organization (WHO) also added PAHs to the list of priority pollutants in both air and water. France, Japan, Germany, Netherlands, Sweden, and Switzerland established emission standards for most of the hazardous air pollutants, including PAHs. The WHO and the Netherlands even established ambient air quality guidelines for PAHs (1.0 ng m^{-3} and 0.5 ng m^{-3}, respectively). These limits are not legally binding, but the common consensus is that these pollutants require maximum reduction or "zero levels" in emissions.

The increased attention to nitro-PAHs is due to their persistence in the environment and the higher mutagenic (2×10^5 times) and carcinogenic (10 times) properties of certain compounds compared to PAHs [14]. Studies have highlighted the important role of nitro compounds, and they emphasised the necessity of improving primary prevention methods to reduce nitrated molecule air pollution. Today, the known sources of these types of compounds are a variety of combustion processes, especially in diesel engines. It was reported that the main contributors to the direct-acting mutagenicity of diesel exhaust particulates were nitro-PAHs, including 1,3-, 1,6- and 1,8-dinitropyrenes, 1-nitropyrene, 4-nitropyrene, and 6-nitrochrysene [15-18]. Therefore, researchers have investigated the chemical and physical properties of particulate matter in diesel exhaust according to the development of diesel technology and emissions regulation; comparisons between the fuel and engine used, such as diesel versus gasoline, light-duty diesel, biodiesel and any so called "clean diesel" have also been reported [19-23].

2.3. Contribution of mobiles sources to atmospheric emissions.

PAHs primarily originate in heavily urbanised or industrialised regions; therefore, the majority of these compounds are anthropogenic. Different sources of PAHs were reviewed and classified according to five main categories (domestic, mobile, industrial, agricultural and natural) [24]. Domestic emissions are predominantly associated with the burning of coal, oil, gas, garbage, or other organic substances such as tobacco or char broiled meat. Mobile sources include the emission from vehicles such as aircraft, shipping vehicles, railways, automobiles, off-road vehicles, and machinery. Industrial sources of PAHs include primary aluminium production, coke production, creosote and wood preservation, waste incineration, cement manufacturing, petrochemical and related industries, bitumen and asphalt industries, rubber tire manufacturing, and commercial heat/power production. Agricultural sources include stubble burning, open burning of moorland heather for regeneration purposes, and open burning of brushwood and straw. With regard to the natural sources of PAHs from terrestrial sources (non-anthropogenic burning of forests, woodlands, and moorlands due to lightning strikes, volcanic eruptions) and cosmic origin, their contributions have been estimated as being negligible to the overall emission of PAHs. However, it is rather difficult to make accurate estimations of PAH emissions because some PAHs are common to a number of these sources, and it is not easy to quantitatively determine how much of a particular PAH comes from a specific source [25]. Additionally, understanding the impact of particular emission sources on the different environmental compartments is crucial for proper risk assessment and risk management. PAHs are always emitted as a mixture, and the relative molecular concentration ratios are considered (often only as an assumption) to be characteristic of a given emission source. Table 2 lists typical diagnostic ratios taken from the literature that show the wide diversity of approaches [26].

Table 2. Diagnostic ratios reviewed from [26].

PAH ratio	Value range	Source	References
Σ(low weight PAHs) / Σ(high weight PAHs)	<1 >1	Pyrogenic Petrogenic	[27]
FLT/(FLT+PYR)	<0.5 >0.5	Petrol emissions Diesel emission	[28]
ANT/(ANT+PHEN)	<0.1 >0.1	Petrogenic pyrogenic	[29]
FLT/(FLT+PYR)	<0.4 0.4-0.5 >0.5	Petrogenic Fossil fuel combustion Grass, wood, coal combustion	[30]
B(a)PYR/(B(a)PYR+CHRYS)	0.2-0.35 >0.35 <0.2 >0.35	Coal combustion Vehicular emissions Petrogenic Combustion	[31] [32]
B(a)PYR/(B(a)PYR+B(e)PYR)	~0.5 >0.5	Fresh particles Photolysis (ageing of particles)	[33]
InPYR/(InPYR+B(ghi)PER)	<0.2 0.2-0.5 >0.5	Petrogenic Petroleum combustion Grass, wood, coal combustion	[32]
2-MeNAPH/PHEN	<1 2-6	Combustion Fossil fuels	[34]
Σ MePHE/PHEN	<1 >1	Petrol combustion Diesel combustion	[35]
B(b)FLT/B(k)FLT	2.5-2.9	Aluminium smelter emissions	
B(a)PYR/B(ghi)PER	<0.6 >0.6	Non traffic emissions Traffic emissions	[36]

Diagnostic ratios should be used with caution; the reactivity of some PAH species with other atmospheric species, such as ozone and/or oxides of nitrogen can change the diagnostic ratio [36-40]. The difference in chemical reactivity, volatility and solubility of PAH species may also introduce bias but to minimise this error, diagnostic ratios obtained from PAHs that have similar physico-chemical properties is mainly used [28]. Regardless, vehicular emissions were shown to be a major source of PAHs, whether the diagnostic ratios were coupled or not with principle component analysis, and irrespective of the region or season involved; moreover, studies indicated that diesel exhausts were the largest source of PAHs and nitro-PAHs, compared with the emissions of the other vehicles [39-42].

3. DIESEL EXHAUST COLLECTION, EXTRACTION AND CHEMICAL ANALYSIS

Diesel exhaust contains not only volatile species, which constitute the gaseous phase, but also solid agglomerates of spherical primary particles of approximately 15-30 nm diameter. These agglomerates are larger particles, in the range 60-200 nm (with a lognormal size distribution), which are produced by an accumulation mode corresponding to a coagulation of the primary particles. The agglomerates are made of solids (elemental carbon and metallic ashes) mixed with condensates and adsorbed material (including organic hydrocarbons). The accumulation mode can be accompanied by a nucleation mode, which consists of smaller particles (which are called ultra-fine particles of 10-20 nm) where the organic material is condensed on primary inorganic nuclei (sulphuric acid) [43]. It has been demonstrated that the ultra-fine particles are more easily volatilised, depending on temperature and dilution conditions, because they are mostly composed of volatile condensates and contain little solid material [44]. Therefore, discriminating between the gaseous species and the chemical species present in the solid state in diesel exhausts remains difficult because their formation processes depend on the sampling train design, sampling location, temperature, humidity and dilution rates of exhausts, and also on after-treatment devices.

3.1. Exhaust sampling and collection for pahs analysis

Numerous instruments are used to monitor and characterise particle-bound or gas-phase PAHs in diesel tailpipe emissions. To obtain reliable and reproducible measurements, it is necessary to properly design the sampling system and to choose adequate materials for trapping the chemical substances to be measured (filters, sorbents, etc.). The majority of experiments are performed on chassis dynamometers under predetermined engine load; a heated hose transfers the exhaust from the tailpipe to full flow exhaust dilution tunnels, and therefore, the exhaust gases are collected at low temperature after being mixed and homogenised with the ambient air (with moderate dilution ratios ranging from 6 to 14) in the dilution tunnel, which is connected to a constant volume sampling system (CVS technique) [45, 46]. Chassis dynamometers can also be equipped with dilution systems such as critical-flow Venturi dilution tunnels [47], ejection dilutors [48], or rotating disk diluters [49], but undiluted exhaust can also be directly collected through a sampling probe inserted inside the tailpipe, which is a less conventional method[50].

3. Diesel exhaust collection, extraction and chemical analysis

3.1.1. Collection of the gas phase

The supports used for trapping diesel gaseous exhaust are often the same as those used for air sampling; polyurethane foam plugs (PUF) of appropriate density can trap semivolatile contaminants in aerosols without creating excessive back pressure, and they are typically used to collect the atmospheric gaseous phase [51]; however, sorbents such as XAD-2 resins can also be used to trap the air gas fraction [52]. XAD resins are nonionic macroporous polystyrene-divinylbenzene beads with macroreticular porosity and high surface areas, which give them their adsorptive characteristics for non-polar hydrophobic compounds such as PAHs or their derivatives. Therefore, we can also find these materials for the collection of semi-volatile PAHs and nitro-PAHs in the gas phase of diesel exhaust, because they are inexpensive and easy to handle, store and transport. Different adsorbent resins, such as XAD-2 or XAD-16, coupled to PUF cartridges (making "sandwiches" cartridges) can be used to collect PAHs in the gaseous phase [53-54]. The gaseous PAHs can also be collected using an annular denuder coated with different sorbents: the first stage is coated with XAD-4 resin, the second stage collects the particulate matter and the third stage is composed of a "sandwich" of polyurethane foam plugs and XAD-4 resin, to assess the volatilisation of PAHs from particles [55].

As previously mentioned, the vast majority of diesel exhaust is sampled in a dilution tunnel at low temperatures, where the condensation of the gaseous phase on solid particles is dominant. Additionally, operating at dilution ratios of approximately 10 is far from perfect because typical atmospheric dilution ratios are between 500 and 1000. A consequence is that the nucleation mode is favoured in the dilution tunnels where the saturation ratios are higher than those during atmospheric dilution, and the gaseous species coming out of the tailpipe are generally impoverished. However, one can be interested in the analysis of the gaseous exhaust at high temperature before the appearance of nucleation or condensation. A sampling method that differs from conventional collection methods was developed for this purpose and consists of trapping PAHs in an aqueous solution by gas bubbling and not on a solid sorbent. Therefore, when a particulate filter is incorporated after the diesel engine, the original method for trapping vapour-phase PAHs present in the hot undiluted gaseous exhaust is to absorb them inside an aqueous solution containing various additives, including a cationic surfactant as a solubilising agent [56].

3.1.2. Collection of the particulate phase

Diesel particulate matter (PM) is frequently defined from the material collected on a filter at a temperature of 52°C (or less) after dilution of the diesel exhaust with air. Even if the experiments conducted inside a dilution tunnel do not represent the full range of atmospheric conditions, these sampling conditions are far more common.

With regard to determining particulate PAHs in atmospheric aerosols, many different types of filters can be used and compared for retaining them; filters made of quartz, glass fibres, Teflon coated and nylon can be used [57]. For diesel exhaust, collection systems are also generally composed of glass fibre filters to trap the particulate matter, *e.g.,* Pallflex systems. However Teflon-coated glass fibre filters are preferred over glass fibre filters, because they are more inert to catalysing chemical transformations and are less moisture-sensitive [46].

The sampling procedure using filters to collect particle bound PAHs provides no information about the size of the particulate matter; therefore, a high-flow cascade impactor can be used to replace single filters, which collects particles based on size [58]. Many cascade impactors can separate the particulate matter using different quartz filters placed in each stages into eight size ranges, from particles smaller than 0.4 μm to sizes in the range of 6.6-10.5 μm [21]. Aluminium filters that have silicon grease sprayed onto their surfaces can also be used in each stage of the impactors to prevent the bouncing of the particles during the collection; therefore, the distribution of PAHs based on the particulate sizes can be evaluated [59].

Finally, many studies have focused on particles directly collected inside of a diesel particulate filter (DPF) to understand the effects that an exhaust after-treatment device may have on the formation of particle-bound PAHs; once the DPF filter is weighed for total mass emission control, the particles can be blown-off and recovered for subsequent chemical analysis [60].

3.2. Extraction of pahs from the trapping media

3.2.1. Extraction and purification of trapped gaseous pahs

Vapour-phase PAHs and nitro-PAHs present in the trapped phase are often solvent-extracted using conventional techniques such as an ultrasonic bath or Soxhlet apparatus. A large number of polyurethane foams are Soxhlet extracted with 10% diethyl ether in hexane, as described in US EPA method TO-13.

XAD-2 resins are also often Soxhlet extracted over the course of 8 hours with 120 mL of methylene chloride [53], whereas mixtures of PAHs and nitro-PAHs can be better extracted from XAD-16 resins using 300 mL of methylene chloride:acetonitrile 3:1 (v:v) for 16 hours [21]; the extracts are then concentrated to 1-2 mL using rotary evaporators or under nitrogen flow, and they must be purified before analysis to eliminate matrix interferences. For the cleanup process, the concentrated solutions can be introduced into a silica column that contains anhydrous sodium sulphate to exclude water, and then eluted with 20 mL of methylene chloride:hexane 1:2 (to recover PAHs) and 30 mL of acetone:hexane 1:2 (to better recover nitro-PAHs). Finally, after purification, another concentration step by evaporation is necessary to enhance the concentration of solutes to be analysed. Each of these steps (desorption, evaporation, purification, and evaporation) is time consuming, and large volumes of organic solvents are consumed; additionally, the evaporation concentration steps can be critical for obtaining quantitative results for the most volatile PAHs.

Another method for concentrating and cleaning PAHs (in one step) after their absorption in an aqueous medium is the use of a solid phase extraction process (SPE), which consists of percolating the PAHs solubilising aqueous medium through a short column containing hydrophobic packing material and eluting them directly inside of the analytical apparatus without any losses. This method is less time and solvent consuming [56].

3.2.2. Extraction of pahs from diesel particles

Among all of the solid environmental polluted matrices studied, such as soils, sediments or urban dusts, carbonaceous diesel particulate matter is known to require severe conditions to obtain good recovery yields of the higher molecular weight PAHs [61]. It was even suggested that among all of the natural environmental matrices, diesel PM was the most refractory [62]. Despite this observation, only a few studies have a reliable optimisation of the PAHs extraction step from the carbonaceous matrix.

3.2.2.1. Conventional solvent extraction methods

In a significant number of studies, PAHs are extracted from PM trapped in glass-fibre filters in ultrasonic baths; for example, Karavakalis *et al.* used methylene chloride to extract three times the PAHs, with a total of 80 mL of solvent [45]. Riddle *et al.* used a 1:1 mixture of methylene chloride:hexane for 15 minutes, three times [63]. Fernandes *et al.* mentioned that using an ultrasonic bath for 1 hour with 70 mL of hexane was not as efficient as the Soxhlet extraction of PM with methylene chloride or a 3:1 mixture of methylene chloride:methanol [64]. However, Soxhlet extractions, even if they are more efficient than ultrasonic extractions, are longer; it can take 18 h to 24 h to extract PAHs using Soxhlet extractions [59, 65]. In many cases, even a long Soxhlet extraction does not result in the complete extraction of high weight PAHs and nitrated PAHs from diesel soot.figure 1 shows that increasing the number of Soxhlet extraction cycles, using a classical extraction solvent for PAHs, does not permit the extraction of strongly adsorbed PAHs and nitro-PAHs from diesel soot; however, the aliphatic hydrocarbons are quantitatively extracted. In this example, the studied soot was very poor in the soluble organic fraction (SOF) and the pollutants were strongly adsorbed onto the partially graphitic surface and not absorbed into the condensation layer around the carbonaceous core, which can explain the difficulty of extracting high weight PAHs using classical techniques and solvents. However, more and more engines equipped with DPF result in the production of soot with very few condensed SOF, where the PAHs can be very difficult to extract; therefore, it seems risky to neglect the optimisation of the extraction step.

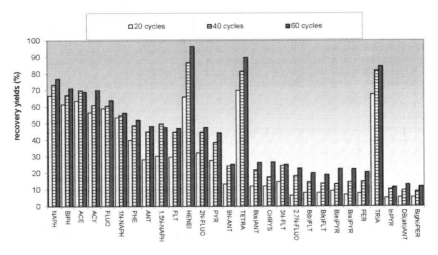

Figure 1. Soxhlet extraction of a mixture of spiked PAHs, nitro-PAHs and n-alkanes from 100 mg of diesel particulate matter, blown-off from a diesel particulate filter after an European NEDC driving cycle test. Extraction conditions: reflux of 120 mL methylene chloride, each cycle representing approximately 8 minutes.

Soxhlet extraction can be enhanced with hot Soxhlet, where heating is also applied to the extraction cavity (unlike conventional Soxhlet), but the temperature must be kept lower than the boiling point of the extracting solvent mixture to keep it in the liquid state; therefore, temperatures are not high, and even if the recoveries are slightly better than using conventional Soxhlet, it remains difficult to quantitatively extract high weight PAHs from poor SOF diesel particles [60].

Finally, before the analytical step, columns containing a mixture of silica gel and deactivated alumina can be used for the fractionation and purification of extracts; the elution with *n*-hexane permits the elimination of aliphatic hydrocarbons and a second elution using 25 mL of a 3:1 mixture of cyclohexane:methylene chloride permits the elution of PAHs and nitro-PAHs [40]. As previously mentioned, all of these extraction and cleanup steps are time and solvent consuming.

3.2.2.2. Recent rapid solvent extraction methods

Other extraction procedures, such as microwave-assisted extraction (MAE), supercritical fluid extraction (SFE) or assisted solvent extraction (ASE) can favourably replace Soxhlet or sonication extractions and yields cleaner extracts with minimal loss of volatile compounds and minimal use of solvents. Although these recent and more effective extraction methods are employed to extract PAHs or nitro-PAHs from solid matrices such as soils, sediments, plants or atmospheric dusts [66-70], they are only little used for the extraction of PAHs from collected diesel soot.

Microwave-assisted extraction uses closed inert vessels that contain the solid matrix and the organic solvents subjected to microwave irradiation (see Figure 2a). An important advantage of the MAE technique is the extraction rate acceleration due to microwave irradiation, resulting in an immediate heating to 120-140°C; therefore, extraction times on the order of a few minutes (approximately 30 minutes) can be obtained compared to a few hours when Soxhlet is used. Solvent volumes are low and range between 15 and 40 mL, depending on the quantity of matrix to be extracted. The solvent mixtures may have a dielectrical constant that is large enough to permit the heating transfer, and typical solvents such as acetonitrile or mixtures of toluene: acetone are employed to extract PAHs from atmospheric particles [71, 72]. In fact, the choice of the nature and the volume of the extracting solvents is especially important for the quantitative extraction of PAHs from diesel particles. Additionally, in the case of diesel soot with a very little SOF fraction, mixtures of drastic solvents such as pyridine: diethylamine or pyridine: acetic acid must be employed to quantitatively extract high weight PAHs or nitro-PAHs [73].

Supercritical fluid extraction exploits the unique properties of a supercritical fluid to more rapidly extract organic analytes from solid matrices, because the supercritical state imparts a great diffusion rate (like in the gaseous state) and a good solvating power (like in the liquid state). Carbon dioxide is the most used supercritical fluid because its critical temperature of 31°C and critical pressure of 74 bar are easily obtained.

Supercritical fluids offer the opportunity to control the solvating power of the extraction fluid by varying the temperature (50 to 150 °C) and the pressure (100 to 400 bars), but also by adding minimal organic modifier (Figure 2b). In fact, the addition of small amounts of cosolvent into the CO_2 is absolutely necessary, not only to increase the PAH solubility in the extractant fluid but also to break the strong interactions between the aromatic retention sites of soot surface and the PAHs and nitro-PAHs: cosolvent mixtures of methylene chloride and toluene can be added to better extract PAHs from diesel soot [74], but other less conventional cosolvent mixtures, containing pyridine, are sometimes required [75]. SFE is an attractive method, as it leads to rapid extractions (less than 45 minutes, combining static and dynamic steps) and generates extracts ready for analysis without the need for additional concentration by means of solvent evaporation, because the extracted solutes are recovered into a very small volume of solvent. Moreover, SFE provides clean extracts because of its higher selectivity when compared to liquid solvent extraction techniques. Consequently, time consuming clean-up steps, which are also a source of analytical error, are not absolutely necessary after SFE extractions [76].

More recently, a method named accelerated solvent extraction (ASE) or pressurised fluid extraction (PFE) was developed. This method uses an organic solvent at a relatively high pressure and temperature to achieve more rapid extractions from solid matrices (Figure 2c) [77]. Elevated temperatures (150-200°C) permit to disrupt the strong solute-matrix interactions, and a high pressure (100-150 bars) forces the extraction solvent into the matrix pores. Less than 20 mL of solvent can be used to extract PAHs and nitro-PAHs from diesel PM; extractions with toluene [61], methylene chloride [16] or less conventional solvents based on pyridine [60] were performed.

Finally, we stress the importance of optimising the PAH extraction step from soot; a chemometric approach can be useful because the influence of several operating variables must be understood. Indeed, the use of an experimental design fully permits the evaluation of not only the effects of each variable but also the interaction effects between the studied parameters, and at last permits the construction of a mathematical model that relates the observed response (PAH recovery yields in %) to the various factors and to their combinations (Figure 3).

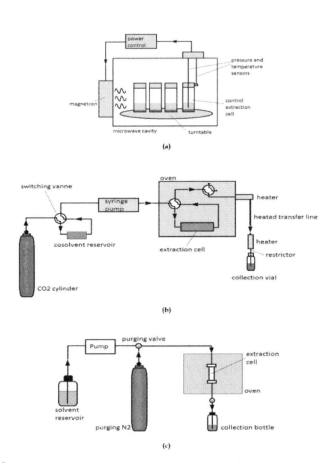

Figure 2. Extraction apparatus (a) Microwave-Assisted Extraction: MAE extraction vessels are placed on a turntable inside an oven and are subjected to microwave irradiations generated by a magnetron. A programmable microcomputer controls and monitors the power, temperature and pressure within the vessel (b) Supercritical Fluid Extraction: SFE incorporates pumps which produce the high pressures required for supercritical CO_2 work. An organic cosolvent can be added to the extraction fluid by a separate module. The extraction cell is placed in an oven which maintains a precise fluid temperature. A static mode (equilibration) and a dynamic mode (with a controlled flow-rate) can be performed successively. A restrictor valve (changing the CO_2 from condensed to gaseous phase) provides precise control over flow-rates (c) Assisted-Solvent Extraction: ASE operates by moving the extraction solvent through an extraction cell which is heated by direct contact of the oven. Extractions can be performed first in static mode; when the extraction is complete, compressed nitrogen moves the heated extracting solvent to the collection bottle.

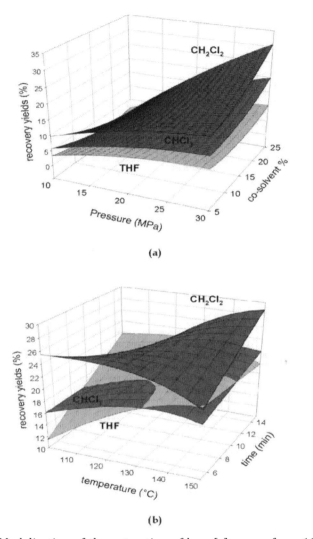

Figure 3. Modelisation of the extraction of benz[a]pyrene from 100 mg of spiked diesel soot, as a function of several influent factors (a) SFE recoveries as a function of pressure and percentage of co-solvent (methylene chloride, chloroform or tetrahydrofuran) added into supercritical CO_2, the temperature being fixed at 75°C and the extraction time at 10 minutes (static phase) and 20 minutes (dynamic phase) (b) ASE recoveries as a function of temperature, extraction time, and nature of the extraction solvent, the pressure being fixed at 100 bars.

3.2.2.3. Extractions without solvent

To avoid the use of organic solvents, other extraction techniques for chemical substances from particulate matter are proposed. Solid phase micro-extraction (SPME) seems to provide an alternative to solvent extraction methods [78]. Head-space SPME methods using polydimethylsiloxane fibres were tested on certified diesel particulate matter; it consisted in equilibrating PAHs sorbed on PM between a saline aqueous phase and the SPME fibre, and then the fibre was thermally desorbed inside the injection port of a gas chromatographer. However, this method was acceptable only for two to four PAHs congeners and was long (4 hours) [79]. Another method is to thermally desorb PAHs from filters or PM at 550°C in a thermogravimetric analyser, then recover them in a sampling bag where the vaporised phase is in equilibrium with a SPME fibre. Thereafter, PAHs are desorbed from the SPME fibre into a gas chromatographer [80]. However, for the quantitative analysis of high weight PAHs, the time necessary to establish the equilibrium (especially at higher concentration levels) is very long and can exceed 15h.

Direct thermal desorption (TD) can also be used for the desorption of PAHs from PM and this technique is generally hyphenated to a gas chromatographer for subsequent analysis. However, this fast extraction technique, which overcomes the main drawbacks of solvent extractions, is more often performed for PAHs sorbed on airborne particulate matter [81-83]; therefore, PAHs are desorbed from air dusts at 300-340°C (without pyrolytic degradation), cryogenically trapped and afterwards released in the gas chromatographer injection port at 325-400°C with a helium flow. Thermal desorption of PAHs from diesel exhaust particles seems to be more difficult and less reproducible, especially for high molecular weight PAHs [84].

Laser desorption-ionisation has also been studied to characterise pollutants immobilised on solid particles, and it offers the advantages of being orders of magnitude more sensitive for the analysis of PAHs and more rapid than solvent extraction. As will be discussed later, laser desorption systems are directly hyphenated to mass spectrometer analysers. Particulate PAHs and nitro-PAHs sorbed on soot particles are generally desorbed at an ionisation wavelength of 193 nm or 266 nm by an UV laser, mixed to a nebulised aerosol and transferred to the ion source region of the mass spectrometer [85]; solid samples to be desorbed can also be directly introduced with a probe into the ion source chamber [86, 87]. It has been described that this energetic UV laser desorption technology is vulnerable to matrix effects, and absolute quantification of PAHs seems to be difficult [88]. Therefore, new methods have appeared where particulate PAHs are deposited onto a probe, are irradiated with a less energetic pulsed Infra-Red laser beam

(1064 nm) and the vaporised molecules are photoionised with UV radiation at 118 nm, so fragmentations are minimised and the sensitivity is higher [89].

3.3. Identification and quantitation of pahs and their derivatives

3.3.1. On-line mass spectrometry analysis

To avoid time consuming sample preparation, we previously mentioned that PAHs can be desorbed from soot particles by IR or UV lasers, and after an ionisation step, they can be directly analysed by mass spectrometry. Generally, time-of-flight (TOF) mass spectrometers are used, and mass peaks ranging from 1 to 5000 g mol^{-1} are displayed. The advantages of this on-line analysis are that very small amounts of PM are required (as small as 10^{-12} to 10^{-11} g) and that the mass range is virtually unlimited [90]. Moreover, even if multiple mass spectra are required to yield a representative spectrum for each chemical species present in the particulate effluent, high acquisition rates allow a significant increase in time resolution and measurements can be performed within minutes [85]. The drawbacks are that only qualitative analyses are assured (identification of compound classes), due to effects of species variations in the small irradiated sample area and to the variability of the laser intensity [87]. Moreover, there is not any selectivity on PAHs isomers on mass spectra, and interferences due to alkylated compounds, carbon clusters, polyynes compounds and photofragments of nitro-PAHs are inevitable in spectra because of the complexity of diesel soot particles. On-line carbon speciation of diesel exhaust particles can also be performed by near-edge X-ray absorption spectroscopy (NEXAFS), but the same drawbacks can be underlined: only fingerprints can be established because PAHs isomers cannot be resolved and quantitative information cannot be proposed [91]. Consequently, a separation technique before the detection instrument seems essential to obtain isomer resolution and quantitation.

3.3.2. Chromatographic analysis

3.3.2.1. Gas chromatography (GC)

The vast majority of chromatographic separations of PAHs and nitro-PAHs in gas chromatography were obtained using classical columns with a length of 30 m, and an internal diameter of 0.25 mm; the open tube is generally coated with a stationary phase composed of cross-linked phenyl (5%) methyl (95%) siloxane (film thickness of 0.25 µm). However, many researchers have used a less polar stationary phase for the separation of

PAHs (100% methyl), and a more polar one for nitro-PAHs (50% phenyl) [92]. Original stationary phases sometimes have to be employed to obtain an isomer resolution of non-priority high weight PAHs, which are nevertheless of considerable interest because they have a high carcinogenicity; smectic-liquid crystalline polysiloxane phases permitted the separation of these PAHs [93]. Another method for a better isomer resolution was to use a longer column, which was 50-60 m long [16]. The most used detector for detecting PAHs or nitro-PAHs after their chromatographic separation is the mass spectrometer, which has the advantages of being more sensitive and providing information for identification, compared to the flame ionisation detector [65]. Quadrupole mass analysers are the most used for PAHs detection, with electron ionization energies of 70 eV. Ion traps can also be used, which provide more accurate structural information (for isomer identification) in the MS-MS mode [63]. With regard to nitro-PAHs, the electron-capture detector (ECD) can be employed, which is a sensitive and selective detector [21]. MS detectors can also be used for the detection of nitro-PAHs, but with different ionisation modes than those applied for PAHs; the most frequent mode is negative chemical ionisation (NICI) with methane as a reagent gas [92], but electron monochromator mass spectrometry (EM-MS) appears to be more selective and specific [94].

It must be noted that the use of programmable temperature vaporisation (PTV) injection systems may enhance the sensitivity by injecting up to 40 µL of samples instead of the 1 µL used for classical split-splitless injectors [69]. However, GC injection systems can also be directly hyphenated to thermodesorption units, which permits on-line PAH thermodesorption, GC separation and MS detection [95], as mentioned in a previous chapter. Finally, it will be necessary to count on comprehensive two dimensional GC techniques (GCxGC) in the future to obtain increased separation power and better sensitivity [96].

3.3.2.2. High performance liquid chromatography (HPLC)

As PAHs are lipophilic compounds, separations are generally achieved using reversed phase liquid chromatography with an apolar octadecyl (C_{18}) column and a mobile phase composed of acetonitrile and water. However, phenyl-modified stationary phases can also be employed for the separation of PAHs and their oxidised derivatives, with methanol and water used as a mobile phase [97]. Even if Ultra-Violet detectors can be used to detect PAHs and nitroPAHs [98], fluorescence detectors are more frequently employed for PAHs because of their greater sensitivity and selectivity, even if one of the sixteen priority PAHs cannot be detected (acenaphtylene) [56]. Nitro-PAHs must be derivatised to be fluorescent (reduction to the amine), and other more appropriate detectors can be employed for their detection, such as chemiluminescence or coulometric detectors [17]. Mass spectrometers have also proven to be useful detectors; PAHs can be analysed using an atmospheric pressure chemical ionisation interface (APCI) between the separation column and a single-quadrupole detection system [97]; atmospheric pressure photoionisation interfaces (APPI) can also be coupled to triple quadrupole mass spectrometers, permitting the analysis of PAHs in the positive ion mode, and nitro-PAHs in the negative ion mode, but also providing structural information about the metabolites [99]. Analytical systems were developed to improve the sensitivity of nitro-PAHs analysis (over chemiluminescence detection) using a two dimensional HPLC procedure (on-line derivatisation and separation), an electrospray ionisation source (ESI) and a triple quadrupole mass spectrometer [100]. Finally, a highly original approach consisting of a hyphenated HPLC and GC-MS resulted in the better analysis of isomers of high weight PAHs from diesel particulate matter [101].

4. CONCLUSION

It is a known fact that polyaromatic hydrocarbons in diesel exhaust are harmful. Therefore, it seems crucial to characterise them in both the gaseous and particulate phases, even if they are not yet regulated pollutants. It has been shown here that a variety of measurement protocols are available and that, even if many regulations in diesel exhaust sampling exist, it is not the case for the whole analytical process. And even for the first step of the measurement, the dilution factor and sampling temperature range are not strictly specified, which can induce great variability in diesel particle mass and composition. For that reason, various conclusions have been drawn about the influence of fuel composition, after-treatment devices, engine load, etc., on PAHs emissions; however, results from different research groups remain difficult to compare and all of the analytical artefacts were not resolved. For example, differentiating between gaseous and particle-bound PAHs is still a problem, because of the desorption or re-adsorption of volatile compounds during sampling. Additionally, neglecting the vapour-phase analysis (as suggested in some studies) also introduces significant error, because the toxic equivalent factors of exhaust are not correctly evaluated. The second step of the analytical process is also, without a doubt, an important source of analytical bias. There is not a universal technique for the extraction of every possible PAH or nitro-PAH from a solid matrix, and all of the described techniques have advantages and disadvantages; for example, high time resolution when coupling the laser desorption directly with the detection is not compatible with isomeric resolution, and thermal desorption remains less reproducible than solvent extraction. With regard to solvent extraction, which is the most employed method for the extraction of PAHs and nitro-PAHs from diesel soot, its optimisation is regrettably widely neglected. Conventional extraction methods such as Soxhlet or ultrasonic extractions, which are generally investigated on diesel PM extractions, and conventional solvents, such as methylene chloride or hexane, are in many cases not sufficient to quantitatively extract high weight PAHs and nitro-PAHs from carbonaceous diesel PM [102]. It is particularly the case for "dry" diesel soot, which has a reduced portion of soluble organic fraction. Indeed, in this case, the sorption is dominated by strong surface adsorption while a high amount of SOF attenuates PAHs adsorption, blocking the energetic sites, and thus a simple phase partitioning dominates [103]. Consequently, the presence on the particles of relevant amounts of SOF helps the solvent extraction process; however, it is now relevant to notice that recent diesel engines that are equipped with oxidation catalysts and particulate filters produce leaner particulates, and as it previously mentioned, more drastic extraction conditions are required [61]. Therefore, in the future, it will be important to pay more attention on the optimisation of the extraction step to obtain quantitative results, especially for the highest weight and more toxic PAHs and nitro-PAHs. Then, new standard reference materials with low SOF, issued from more recent diesel engines, should be produced, characterised and commercialised to validate this important analytical step. Finally, it can be emphasised that studies on PAHs other than the sixteen priority ones, as well as on other oxygenated or sulphured derivatives, could be another important task in the future, considering the fact that some of them are particularly more dangerous. For those studies, comprehensive chromatographic techniques (GCxGC) could help in enhancing the resolution of hundreds of aromatic compounds that can be found on diesel particles.

REFERENCES

1. Q Li, A Wyatt, R. M Kamens, 2009Oxidant generation and toxicity enhancement of aged-diesel exhaust. Atmospheric Environment. 4310371042

2. T. T T Lee, B. -K Characteristics, toxicity, and source apportionment of polycylic aromatic hydrocarbons (PAHs) in road dust of Ulsan, Korea. Chemosphere. 7412451253

3. H Andersson, E Piras, J Demma, B Hellman, E Brittebo, 2009Low levels of the air pollutant 1-nitropyrene induce DNA damage, increased levels of reactive oxygen species and endoplasmic reticulum stress in human endothelial cells. Toxicology. 2625764

4. T Okuda, K Okamoto, S Tanaka, Z Shen, Y Han, Z Huo, 2010Measurement and source identification of polycyclic aromatic hydrocarbons (PAHs) in the aerosol in Xi'an, China, by using automated column chromatography and applying positive matrix factorization (PMF). Science of The Total Environment. 40819091914

5. A Duran, M Carmona, J Monteagudo, 2004Modelling soot and SOF emissions from a diesel engine. Chemosphere. 56209225

6. P Tan, Z Hu, K Deng, J Lu, D Lou, G Wan, 2007Particulate matter emission modelling based on soot and SOF from direct injection diesel engines. Energy Conversion and Management. 48510518

7. D Kittelson, W Watts, J Johnson, J Schauer, D Lawson, 2006On-road and laboratory evaluation of combustion aerosols-Part 2: Summary of spark ignition engine results. Journal of Aerosol Science. 37931949

8. IARC (2010IARC monographs on the evaluation of carcinogenic risks to humans92Some non-heterocyclic polycyclic aromatic hydrocarbons and some related exposures.

9. Environmental Health Criteria EHC n°229 (2003Selected nitro- and nitro oxy-polycyclic aromatic hydrocarbonsFirst draft prepared by Kielhorn J, Wahnschaffe U, Mangelsdorf U. Fraunhofer Institute of Toxicology and Aerosol Research, Hanover, Germany.

10. EFSA (2008Scientific Opinion of the Panel on Contaminants in the Food Chain on a request from the European Commission on Polycyclic Aromatic Hydrocarbons in FoodThe EFSA Journal 7241114

11. Third International Symposium on Chemistry and BiologyMichigan (1979Polynuclear aromatic hydrocarbons, in: Jones P., Leber P. (Eds.), Third International Symposium on Chemistry and Biology: carcinogenesis and mutagenesis, Ann Arbor Science.

12. U. S Epa, North Carolina, USA (1990United States Environmental Protection Agency, Cancer risk from outdoor exposure to air toxics, 1Final Report.

13. IARC (1998International Agency for Research on Cancerlist of IARC Evaluations, polynuclear aromatic hydrocarbons (PAH, Air quality

guidelines for Europe). Copenhagen, WHO Regional Office for Europe, 105117

14. J. L Durant, Busby Jr. W.F, Lafleur A.L, Penman B.W, Crespi C.L (1996Human cell mutagenicity of oxygenated, nitrated and unsubstituted polycyclic aromatic hydrocarbons associated with urban aerosols. Mutation Research. 37134

15. D Traversi, T Schiliro, R Degan, C Pignata, Alessandria., Gilli G (2011Involvement of nitro-compounds in the mutagenicity of urban PM2.5 and PM10 in Turin. Mutation Research. 7265459

16. H. A Bamford, D. Z Bezabeh, M. M Schantz, S. A Wise, J. E Baker, 2003Determination and comparison of nitrated- polycyclic aromatic hydrocarbons measured in air and diesel particulate reference materials, Chemosphere. 50575587

17. X Yang, K Igarashi, N Tang, J. M Lin, W Wang, T Kameda, A Toriba, K Hayakawa, 2010Indirect- and direct-acting mutagenicity of diesel, coal and wood burning-derived particulates and contribution of polycyclic aromatic hydrocarbons and nitro-polycyclic aromatic hydrocarbons. Mutation Research. 6952934

18. R Taga, N Tang, T Hattori, K Tamura, S Sakaic, Toriba., Kizu R, Hayakawa K (2005Direct-acting mutagenicity of extracts of coal burning-derived particulates and contribution of nitro-polycyclic aromatic hydrocarbons. Mutation Research. 5819195

19. M. M Maricq, 2007Chemical characterization of particulate emissions from diesel engines: a review. Aerosol Science. 3810791118

20. C-T Chang, B-Y Chen, 2008Toxicity assessment of volatile organic compounds and polycyclic aromatic hydrocarbons in motorcycle exhaust. Journal of Hazardous Materials. 15312621269

21. H-L Chiang, Y-M Lai, S-Y Chang, 2012Pollutant constituents of exhaust emitted from light-duty diesel vehicles. Atmospheric Environment. 47399406

22. L Turrio-baldassarri, C. L Battistelli, L Conti, R Crebelli, B De Berardis, A. L Iamiceli, M Gambino, S Iannaccone, 2004Emission comparison of urban bus engine fueled with diesel oil and 'biodiesel' blend. Science of the Total Environment. 327147162

23. C Bergvall, R Westerholm, 2009Determination of highly carcinogenic dibenzopyrene isomers in particulate emissions from two diesel and two gasoline-fuelled light-duty vehicles. Atmospheric Environment. 4338833890

24. K Ravindra, S Ranjeet, R Van Grieken, 2008Review: Atmospheric polycyclic aromatic hydrocarbons: source attribution, emissions factors and regulation. Atmospheric Environment. 4228952921

25. M. M. R Mostert, G. A Ayoko, S Kokot, 2010Application of chemometrics to analysis of soil pollutants. Trends in Analytical Chemistry. 29430435

26. M Tobiszewski, J Namiesnik, 2012Review: PAH diagnostic ratios for the identification of pollution emission Sources. Environmental Pollution. 162110119

27. X. L Zhang, S Tao, W. X Liu, Y Yang, Q Zuo, S. Z Liu, 2005Source diagnostics of polycyclic aromatic hydrocarbons based on species ratios: a multimedia approach. Environmental Science and Technology. 3991099114

28. K Ravindra, E Wauters, R Van Grieken, 2008Variation in particulate PAHs levels and their relation with the trans-boundary movement of the air masses. Science of the Total Environment. 396100110

29. C Pies, B Hoffmann, J Petrowsky, Y Yang, T. A Ternes, T Hofmann, 2008Characterization and source identification of polycyclic aromatic hydrocarbons (PAHs) in river bank soils. Chemosphere. 7215941601

30. R. J De La Torre-roche, W-Y Lee, S. I Campos-díaz, 2009Soil-borne polycyclic aromatic hydrocarbons in El Paso, Texas: analysis of a potential problem in the United States/Mexico border region. Journal of Hazardous Materials. 163946958

31. M Akyüz, H Çabuk, 2010Gas particle partitioning and seasonal variation of polycyclic aromatic hydrocarbons in the atmosphere of Zonguldak, Turkey. Science of the Total Environment. 40855505558

32. M. B Yunker, R. W Macdonald, R Vingarzan, R. H Mitchell, D Goyette, S Sylvestre, 2002PAHs in the Fraser River basin: a critical appraisal of PAH ratios as indicators of PAH source and composition. Organic Geochemistry. 33489515

33. C Oliveira, N Martins, J Tavares, C Pio, M Cerqueira, M Matos, H Silva, C Oliveira, F Camoes, 2011Size distribution of polycyclic aromatic hydrocarbons in a roadway tunnel in Lisbon, Portugal. Chemosphere. 8315881596

34. K Opuene, I. E Agbozu, O. O Adegboro, 2009A critical appraisal of PAH indices as indicators of PAH source and composition in Elelenwo Creek, southern Nigeria. Environmentalist. 294755

References

35. M. S Callén, M. T De La Cruz, J. M López, A. M Mastral, 2011PAH in airborne particulate matter. Carcinogenic character of PM10 samples and assessment of the energy generation impact. Fuel Processing and. 92176182

36. A Katsoyiannis, E Terzi, Q-Y Cai, 2007On the use of PAH molecular diagnostic ratios in sewage sludge for the understanding of the PAH sources. Is this use appropriate? Chemosphere. 6913371339

37. E Galarneau, 2008Source specificity and atmospheric processing of airborne PAHs: implications for source apportionment. Atmospheric Environment. 4281398149

38. W Zhang, S Zhang, C Wan, D Yue, Y Ye, X Wang, 2008Source diagnostics of polycyclic aromatic hydrocarbons in urban road runoff, dust, rain and canopy through fall. Environmental Pollution. 153594601

39. K Saarnio, M Sillanpää, R Hillamo, E Sandell, A. R Pennanen, R. O Solanen, 2008Polycyclic aromatic hydrocarbons in size-segregated particular matter from six urban sites in Europe. Atmospheric Environment. 4290879097

40. G Karavalakis, G Fontaras, D Ampatzoglou, M Kousoulidou, S Stournas, Z Samaras, E Bakeas, 2010Effects of low concentration biodiesel blends application on modern passenger cars. Part 3: Impact on PAH, nitro PAH and oxy-PAH emissions. Environmental Pollution. 15815841594

41. M Mari, R. M Harrison, M Schuhmacher, J. L Domingo, S Pongpiachan, 2010Interferences over the sources and processes affecting polycyclic aromatic hydrocarbons in the atmosphere derived from measured data. Science of the Total Environment. 40823872393

42. K Magara-gomez, M. R Olson, T Okuda, K. A Walz, 2012Sensitivity of hazardous air pollutant emissions to the combustion of blends of petroleum diesel and biodiesel fuel. Atmospheric Environment. 50307313

43. H Burtscher, 2005Physical characterization of particulate emissions from diesel engines: a review. Aerosol Science. 36896932

44. N Bukowiecki, D Kittelson, W Watts, H Burtscher, E Weingartner, U Baltensperger, 2002Real-time characterization of ultrafine and accumulation mode particles in ambient combustion aerosols. Journal of Aerosol Science. 3311391154

45. G Karavalakis, E Bakeas, G Fontaras, S Stournas, 2011Effect of biodiesel origin on regulated and particle-bound PAH (polycyclic aromatic hydrocarbon) emissions from a Euro 4 passenger car. Energy. 3653285337

46. P Pakbin, Z Ning, J. J Schauer, C Sioutas, 2009Characterization of Particle Bound Organic Carbon from Diesel Vehicles Equipped with Advanced Emission Control Technologies. Environmental Science & Technology. 4346794686

47. Z. G Liu, D. R Berg, T. A Swor, J. J Schauer, 2008Comparative Analysis on the Effects of Diesel Particulate Filter and Selective Catalytic Reduction Systems on a Wide Spectrum of Chemical Species Emissions. Environmental Science & Technology. 4260806085

48. C He, Y Ge, J Tan, K You, X Han, J Wang, 2010Characteristics of polycyclic aromatic hydrocarbons emissions of diesel engine fueled with biodiesel and diesel. Fuel. 8920402046

49. L-H Young, Liou Y-J; Cheng M-T, Lu J-H, Yang H-H, Tsai Y.I, Wang L-C, Chen C-B, Lai J-S (2012Effects of biodiesel, engine load and diesel particulate filter on nonvolatile particle number size distributions in heavy-duty diesel engine exhaust. Journal of Hazardous Materials. 200282289

50. N. V Heeb, P Schmid, M Kohler, E Gujer, M Zennegg, D Wenger, A Wichser, A Ulrich, U Gfeller, P Honegger, K Zeyer, L Emmenegger, J-L Petermann, J Czerwinski, T Mosimann, M Kasper, A Mayer, 2008Secondary Effects of Catalytic Diesel Particulate Filters: Conversion of PAHs versus Formation of Nitro-PAHs. Environmental Science & Technology. 4237733779

51. M Dimashki, S Harrad, R. M Harrison, 2000Measurements of nitro-PAH in the atmospheres of two cities. Atmospheric Environment. 3424592469

52. F Esen, Y Tasdemir, N Vardar, 2008Atmospheric concentrations of PAHs, their possible sources and gas-to-particle partitioning at a residential site of Bursa, Turkey. Atmospheric Research. 88243255

53. R De Abrantes, J. V De Assunçao, C. R Pesquero, 2004Emission of polycyclic aromatic hydrocarbons from light-duty diesel vehicles exhaust. Atmos. Environ. 3816311640

54. H-H Yang, L-T Hsieh, H-C Liu, H-H Mi, 2005Polycyclic aromatic hydrocarbon emissions from motorcycles. Atmospheric Environment. 391725

References

55. B Zielinska, J Sagebiel, W. P Arnott, C. F Rogers, K. E Kelly, D. A Wagner, J. S Lighty, A. F Sarofim, G Palmer, 2004Phase and Size Distribution of Polycyclic Aromatic Hydrocarbons in Diesel and Gasoline Vehicle Emissions. Environmental Science & Technology. 3825572567

56. F Portet-koltalo, D Preterre, F Dionnet, 2011A new analytical methodology for a fast evaluation of semi-volatile polycyclic aromatic hydrocarbons in the vapor phase downstream of a diesel engine particulate filter. Journal of Chromatography A. 1218981989

57. E Borras, L. A Tortajada-genaro, 2007Characterization of polycyclic aromatic hydrocarbons in atmospheric aerosols by gas-chromatography-mass spectrometry. Analytica Chimica Acta. 583266276

58. W. W Song, K. B He, J. X Wang, X. T Wang, X. Y Shi, C Yu, W. M Chen, L Zheng, 2011Emissions of EC, OC, and PAHs from Cottonseed Oil Biodiesel in a Heavy-Duty Diesel Engine. Environmental Science & Technology. 4566836689

59. Y-C Lin, C-H Tsai, C-R Yang, C. J Wu, T-Y Wu, G-P Chang-chien, 2008Effects on aerosol size distribution of polycyclic aromatic hydrocarbons from the heavy-duty diesel generator fueled with feedstock palm-biodiesel blends. Atmospheric Environment. 4266796688

60. K Oukebdane, F Portet-koltalo, N Machour, F Dionnet, P. L Desbene, 2010Comparison of hot Soxhlet and accelerated solvent extractions with microwave and supercritical fluid extractions for the determination of polycyclic aromatic hydrocarbons and nitrated derivatives strongly adsorbed on soot collected inside a diesel particulate filter. Talanta. 82227236

61. L Turrio-baldassarri, C. L Battistelli, A. L Iamiceli, 2003Evaluation of the efficiency of extraction of PAHs from diesel particulate matter with pressurized solvents. Analytical and Bioanalytical Chemistry. 375589595

62. B. A Benner, 1998Summarizing the Effectiveness of Supercritical Fluid Extraction of Polycyclic Aromatic Hydrocarbons from Natural Matrix Environmental Samples. Analytical Chemistry. 7045944601

63. S. G Riddle, C. A Jakober, M. A Robert, T. M Cahill, M. J Charles, M. J Kleeman, 2007Large PAHs detected in fine particulate matter emitted from light-duty gasoline vehicles. Atmospheric Environment. 4186588668

64. M. B Fernandes, P Brooks, 2003Characterization of carbonaceous combustion residues: II. Nonpolar organic compounds. Chemosphere. 53447458
65. N. Y Rojas, H. A Milquez, H Sarmiento, 2011Characterizing priority polycyclic aromatic hydrocarbons (PAH) in particulate matter from diesel and palm oil-based biodiesel B15 combustion. Atmospheric Environment. 4561586162
66. S. B Hawthorne, C. B Grabanski, E Martin, D. J Miller, 2000Comparisons of Soxhlet extraction, pressurized liquid extraction, supercritical fluid extraction and subcritical water extraction for environmental solids: recovery, selectivity and effects on sample matrix. Journal of Chromatography A. 892421433
67. F Priego-capote, J Luque-garcia, Luque de Castro M (2003Automated fast extraction of nitrated polycyclic aromatic hydrocarbons from soil by focused microwave-assisted Soxhlet extraction prior to gas chromatography-electron-capture detection. Journal of Chromatography A. 994159167
68. V Yusa, G Quintas, O Pardo, A Pastor, M Guardia, 2006Determination of PAHs in airborne particles by accelerated solvent extraction and large-volume injection-gas chromatography-mass spectrometry. Talanta. 69807815
69. P Bruno, M Caselli, G De Gennaro, M Tutino, 2007Determination of polycyclic aromatic hydrocarbons (PAHs) in particulate matter collected with low volume samplers. Talanta. 7213571361
70. N Itoh, A Fushimi, T Yarita, Y Aoyagi, M Numata, 2011Accurate quantification of polycyclic aromatic hydrocarbons in dust samples using microwave-assisted solvent extraction combined with isotope-dilution mass spectrometry. Analytica Chimica Acta. 6994956
71. Y. Y Shu, S. Y Tey, D. K Wu, 2003Analysis of polycyclic aromatic hydrocarbons in airborne particles using open-vessel focused microwave-assisted extraction. Analytica Chimica Acta. 49599108
72. D Castro, K Slezakova, C Delerue-matos, M. C Alvim-ferraz, S Morais, M. C Pereira, 2011Polycyclic aromatic hydrocarbons in gas and particulate phases of indoor environments influenced by tobacco smoke: Levels, phase distributions, and health risks. Atmospheric Environment. 4517991808
73. F Portet-koltalo, K Oukebdane, F Dionnet, P. L Desbene, 2008Optimisation of the extraction of polycyclic aromatic hydrocarbons and their nitrated derivatives from diesel particulate

matter using microwave-assisted extraction. Analytical and Bioanalytical Chemistry. 390389398

74. C Jones, A Chughtai, B Murugaverl, D Smith, 2004Effects of air/fuel combustion ratio on the polycyclic aromatic hydrocarbon content of carbonaceous soots from selected fuels. Carbon. 4224712484

75. F Portet-koltalo, K Oukebdane, F Dionnet, P. L Desbene, 2009Optimisation of supercritical fluid extraction of polycyclic aromatic hydrocarbons and their nitrated derivatives adsorbed on highly sorptive diesel particulate matter. Analytica Chimica Acta. 6514856

76. S Bowadt, S. B Hawthorne, 1995Supercritical fluid extraction in environmental analysis. Journal of Chromatography A. 703549571

77. M Ryno, L Rantanen, E Papaioannou, A. G Konstandopoulos, T Koskentalo, K Kavela, 2006Comparison of pressurized fluid extraction, Soxhlet extraction and sonication for the determination of polycyclic aromatic hydrocarbons in urban air and diesel exhaust particulate matter. Journal of Environmental Monitoring. 8488493

78. B Tang, U Isacsson, 2008Analysis of Mono- and Polycyclic Aromatic Hydrocarbons Using Solid-Phase Microextraction: State-of-the-Art. Energy & Fuels. 2214251438

79. J. M Vaz, 2003Screening direct analysis of PAHS in atmospheric particulate matter with SPME. Talanta. 60687693

80. R Ballesteros, J Hernandez, L Lyons, 2009Determination of PAHs in diesel particulate matter using thermal extraction and solid phase micro-extraction. Atmospheric Environment. 43655662

81. B. L Van Drooge, I Nikolova, P. P Ballesta, 2009Thermal desorption gas chromatography-mass spectrometry as an enhanced method for the quantification of polycyclic aromatic hydrocarbons from ambient air particulate matter. Journal of Chromatography A. 121640304039

82. J Gil-molto, M Varea, N Galindo, J Crespo, 2009Application of an automatic thermal desorption-gas chromatography-mass spectrometry system for the analysis of polycyclic aromatic hydrocarbons in airborne particulate matter. Journal of Chromatography A. 121612851289

83. T Ancelet, P. K Davy, W. J Trompetter, A Markwitz, D. C Weatherburn, 2011Carbonaceous aerosols in an urban tunnel. Atmospheric Environment. 4544634469

84. R. J Lavrich, M. D Hays, 2007Validation Studies of Thermal Extraction-GC/MS Applied to Source Emissions Aerosols. 1. Semivolatile Analyte- Nonvolatile Matrix Interactions. Analytical Chemistry. 7936353645

85. M Kalberer, B. D Morrical, M Sax, R Zenobi, 2002Picogram Quantitation of Polycyclic Aromatic Hydrocarbons Adsorbed on Aerosol Particles by Two-Step Laser Mass Spectrometry. Analytical Chemistry. 7434923497

86. R. N Dotter, C. H Smith, M. K Young, P. B Kelly, A. D Jones, E. M Mccauley, D. P. Y Chang, 1996Laser Desorption/Ionization Time-of-Flight Mass Spectrometry of Nitrated Polycyclic Aromatic Hydrocarbons. Analytical Chemistry. 6823192324

87. R. A Dobbins, R. A Fletcher, B. A Benner, S Hoeft, 2006Polycyclic aromatic hydrocarbons in flames, in diesel fuels, and in diesel emissions. Combustion and Flame. 144773781

88. M Bente, M Sklorz, T Streibel, R Zimmermann, 2008Online Laser Desorption-Multiphoton Postionization Mass Spectrometry of Individual Aerosol Particles: Molecular Source Indicators for Particles Emitted from Different Traffic-Related and Wood Combustion Sources. Analytical Chemistry. 8089919004

89. T Ferge, F Muhlberger, R Zimmermann, 2005Application of Infrared Laser Desorption Vacuum-UV Single-Photon Ionization Mass Spectrometry for Analysis of Organic Compounds from Particulate Matter Filter Samples. Analytical Chemistry. 7745284538

90. A Faccinetto, P Desgroux, M Ziskind, E Therssen, C Focsa, 2011High-sensitivity detection of polycyclic aromatic hydrocarbons adsorbed onto soot particles using laser desorption/laser ionization/time-of-flight mass spectrometry: An approach to studying the soot inception process in low-pressure flames. Combustion and Flame. 158227239

91. A Braun, B. S Mun, F. E Huggins, G. P Huffman, 2006Carbon Speciation of Diesel Exhaust and Urban Particulate Matter NIST Standard Reference Materials with C(1s) NEXAFS Spectroscopy. Environmental Science & Technology. 41173178

92. Y Kawanaka, K Sakamoto, N Wang, S-J Yun, 2007Simple and sensitive method for determination of nitrated polycyclic aromatic hydrocarbons in diesel exhaust particles by gas chromatography-negative ion chemical ionisation tandem mass spectrometry. Journal of Chromatography A. 1163312317

References

93. P Schubert, M. M Schantz, L. C Sander, S. A Wise, 2003Determination of Polycyclic Aromatic Hydrocarbons with Molecular Weight 300 and 302 in Environmental-Matrix Standard Reference Materials by Gas Chromatography/Mass Spectrometry. Analytical Chemistry. 75234246

94. M. A Ratcliff, A. J Dane, A Williams, J Ireland, J Luecke, R. L Mccormick, K. J Voorhees, 2010Diesel Particle Filter and Fuel Effects on Heavy-Duty Diesel Engine Emissions. Environmental Science & Technology. 4483438349

95. M Bente, M Sklorz, T Streibel, R Zimmermann, 2009Thermal Desorption-Multiphoton Ionization Time-of-Flight Mass Spectrometry of Individual Aerosol Particles: A Simplified Approach for Online Single-Particle Analysis of Polycyclic Aromatic Hydrocarbons and Their Derivatives. Analytical Chemistry. 8125252536

96. M. Z Ozel, J. F Hamilton, A. C Lewis, 2011New Sensitive and Quantitative Analysis Method for Organic Nitrogen Compounds in Urban Aerosol Samples. Environmental Science & Technology. 4514971505

97. T Letzel, U Poschl, R Wissiack, E Rosenberg, M Grasserbauer, R Niessner, 2001Phenyl-Modified Reversed-Phase Liquid Chromatography Coupled to Atmospheric Pressure Chemical Ionization Mass Spectrometry: A Universal Method for the Analysis of Partially Oxidized Aromatic Hydrocarbons. Analytical Chemistry. 7316341645

98. I El Haddad, N Marchand, J Dron, B Temime-roussel, E Quivet, H Wortham, J. L Jaffrezo, C Baduel, D Voisin, J. L Besombes, G Gille, 2009Comprehensive primary particulate organic characterization of vehicular exhaust emissions in France. Atmospheric Environment. 4361906198

99. C Hutzler, A Luch, J. G Filser, 2011Analysis of carcinogenic polycyclic aromatic hydrocarbons in complex environmental mixtures by LC-APPI-MS/MS. Analytica Chimica Acta. 702218224

100. J. P Miller-schulze, M Paulsen, A Toriba, K Hayakawa, C. D Simpson, 2007Analysis of 1-nitropyrene in air particulate matter standard reference materials by using two-dimensional high performance liquid chromatography with online reduction and tandem mass spectrometry detection. Journal of Chromatography A. 1167154160

101. C Bergvall, R Westerholm, 2008Determination of 252-302 Da and tentative identification of 316-376 Da polycyclic aromatic hydrocarbons in Standard Reference Materials 1649a Urban Dust and 1650b and 2975 Diesel Particulate Matter by accelerated solvent extraction-HPLC-GC-MS. Analytical and Bioanalytical Chemistry. 39122352248

102. M. T. O Jonker, A. A Koelmans, 2002Extraction of Polycyclic Aromatic Hydrocarbons from Soot and Sediment: Solvent Evaluation and Implications for Sorption Mechanism. Environmental Science & Technology. 3641074113

103. S Endo, P Grathwohl, S. B Haderlein, T. C Schmidt, 2009Effects of Native Organic Material and Water on Sorption Properties of Reference Diesel Soot. Environmental Science & Technology. 4331873193

CHAPTER 3

A Model-Free Diagnosis Approach for Intake Leakage Detection and Characterization in Diesel Engines

Ghaleb Hoblos* and Mourad Benkaci

ESIGELEC-IRSEEM, Avenue Galilée, 76801 Saint Etienne du Rouvray, France

ABSTRACT

Feature selection is an essential step for data classification used in fault detection and diagnosis processes. In this work, a new approach is proposed, which combines a feature selection algorithm and a neural network tool for leak detection and characterization tasks in diesel engine air paths. The Chi square classifier is used as the feature selection algorithm and the neural network based on Levenberg-Marquardt is used in system behavior modeling. The obtained neural network is used for leak detection and characterization. The model is learned and validated using data generated by xMOD. This tool is used again for testing. The effectiveness of the proposed approach is illustrated in simulation when the system operates on a low speed/load and the considered leak affecting the air path is very small.

Keywords: leak detection; automotive diagnosis; feature selection; neural data classification; diesel air path

1. INTRODUCTION

In order to reduce air pollution caused by automobile engines, several standards have been introduced. The first standard was proposed by the California Air Resources Board in 1970. Since 1993, marked by the introduction of the Kyoto Protocol, the European anti-pollution standards have become more stringent where the authorized emissions of a diesel vehicle have been decreased from (NOx = nil, CO = 2720, HC + NOx = 970, PM = 140) in Euro1 standard to (NOx = 80, CO = 500, HC + NOx = 170, PM = 5) in Euro6 standard. Typically, each fault that increases the emission level must be detected and isolated. The leaks in the intake canal are among the most difficult faults to manage. Several works have been devoted to fault detection and isolation (FDI) in automotive applications [1,2,3,4]. According to the above-mentioned references, process knowledge used in the form of a mathematical process model enables model-based FDI approaches to enhance the performance of the engine diagnosis systems. Nyberg [1,2] used structured hypothesis testing for additive and multiplicative faults in production engines. A real-time observing method which uses observers that estimate unknown fault-parameters was discussed in [3]. In [4], structured parity equations are used to detect and isolate malfunctions in sensors and actuators.

The ability of neural networks to approximate a nonlinear function makes them one of the best tools for fault detection and isolation. By exploring these artificial intelligence techniques, researchers developed other classes of fault detection and isolation algorithms. In [5], Isermann discussed the superior features of neural networks in fault classification and recognition. Sorsa and Costin [6] show the capabilities of supervised neural networks such as Multilayers Perceptron (MLP) and Radial Basis Function (RBF) to perform good and effective fault detection and isolation tasks. Another work using MLP network is presented in [7]. The RBF was used in on-board fault diagnosis for the air path of spark ignition engines [8]. The leakage problem of gasoline engines is dealt with in [9] where the neural network based on the steepest-descent method combined with a back propagation algorithm is developed to train three detection systems.

Because of the increased complexity of today's engines, which are characterized by a large number of sensors, the reduction of the acquired data becomes essential.

Feature selection is one of important steps before beginning a data classification task, especially when this task is dedicated to fault detection and isolation. It refers to the problem of selecting the input features that are most predictive for a given outcome. The feature selection problems are found in all supervised and unsupervised machine learning which include classification, regression, time-series, prediction and clustering. The feature selection tasks try to achieve three main goals: reduce the cost of extracting features, improve the classification accuracy and enhance the reliability of the estimated performances.

There are many works that use the feature selection algorithm for fault detection and diagnosis. In reference [10], a new approach for intrusion detection and diagnosis is proposed. It combines the sequential backward floating search which selects the pertinent features with the fuzzy ARTMAP [11] classification used for detection and diagnosis of attacks. In this work, the vigilance parameter of fuzzy ARTMAP is fixed using a Genetic Algorithm (GA). In [12], the authors propose using a new optimization algorithm in the feature selection procedure. This algorithm is based on a modified binary particle swarm optimization with mutation combined with support vector machine classification. In [13] an improved distance evaluation technique is used to select the optimal features and the importance of features in diagnosis processes is pondered by weight. In reference [14], the authors use the decision tree to identify the best features in the classification task. They use a proximal support vector machine characterized by its capability to efficiently classify the faults in Roller Bearing systems. In [15], the authors perform a diagnosis of induction motors using pattern recognition methods. They use a sequential backward algorithm in order to select the most relevant features. In the previous work, the classification is assured by the k-nearest neighbors rule including reject options. The authors of [16] propose to incorporate a GA with Fisher Discriminant Analysis (FDA) in the key variables identification procedure. The GA is used to select the features that optimize the FDA classification success rate. This approach is applied to the data generated by the Tennessee Eastman Process (TEP) simulator. A new method for feature selection based on mutual information for fault detection and identification is proposed in [17]. Their algorithm is based on two principle stages: the variables are sorted based on their shared mutual information with the class variable and secondly the more informative variables are chosen by taking into account the classification error rate. Once more, the approach is applied in the TEP (Tennessee Eastman Process) simulator. In [18], the authors use a recursive feature elimination to select key variables using Support Vector Machines (SVM). The SVM is combined with time lags incorporated before every classification step. Finally, the number of relatively important variables determined by each classifier is basically determined by 10-fold cross-validation. Wang [19] introduces a neural network approach to vibration feature selection in mechanical systems fault detection. He proposes an artificial intelligence methodology for mechanical fault detection using vibration data, which includes intelligent feature optimization. He uses a back-propagation neural network twice, the first one for feature selection and the second for fault detection.

In this paper, a new methodology dealing with the problem of detecting and characterizing small leaks in diesel air paths is developed. To achieve this goal, a new scheme based on a neural network technique is proposed. The nominal mode (without leaks) and leakage mode corresponding to several diameters of leaks were trained using a Levenberg-Marquardt algorithm. Before using the acquired data, a feature selection task is proposed in order to reduce the complexity of the problem. The main challenge of the proposed approach is the use of selected sensors leading to a reduced cost. The data for the considered modes are generated using the xMOD platform which will be described later.

The paper is organized in this way. First, the considered problem is presented in Section 2. Section 3 describes the proposed approach in detail. A brief description of neural networks based on the steepest-descent and Gauss-Newton methods is given, and the main detection and characterization scheme is illustrated. After a brief description of the MOD tool used in engine data collecting, Section 4 gives some results obtained using our approach. Then, these results are discussed and commented in order to illustrate the effectiveness of leakage detection and characterization.

2. PROBLEM STATEMENT

Over the past several years, anti-pollution standards have become more stringent and then the constraints for the automotive industry have also become very complex. The main objective of these standards is to reduce the emission levels of cars. In the case of diesel engines, there are several pollutants: carbon monoxide, unburned hydrocarbons, nitrogen oxides (NOx) and diesel particulate matter. Usually, the emission level proportionally increases with the appearance of faults in diesel engines, more precisely in diesel air paths. These faults can be due to sensor failures, actuator failures or system degradation. In this paper, the latter failure class is considered. More precisely, the leakage detection and characterization in diesel air paths is studied. This failure can cause multiple non-desired system behaviors. In addition to the high emission levels, this failure causes multiple non-desired effects such as:

- Changes in the operating points of the air path subsystems,
- Incomplete combustion in cylinders,
- Appearance of smoke and the reduction of performance.

Often, this type of failure can be confused with the two other types of faults, *i.e.*, sensors or actuators; consequently, it is very important to distinguish this fault from others.

In addition to the main objective of this paper, the feature selection problem is considered. It is well known that today's vehicles are characterized by increased complexity due to the growing number of embedded sensors. Consequently, the use of selected subsets of sensor data which are in correlation with the considered problem is widely desired in such applications.

In this work, our main objective is to detect and characterize air leaks in diesel air paths regardless of their diameters. Before performing leak detection and characterization, we carry out a feature selection in order to reduce the data complexity.

It is important to specify that, for this application, small leaks are hidden and are very difficult to detect because of the phenomenon of the non-solicitation of the system.

3. PROPOSED APPROACH

Nowadays, the neural network is an essential tool used in many research activities for complex industrial systems. An advantage of using neural networks to detect system faults is that they can interpret the measurement data. Indeed, a neural network has both the ability to generalize an obtained model and, to apply the associative property to the available memory. The error tolerance, characterizing the neural network, effectively deals with the errors of the model. In addition, it can perform nonlinear mapping and also learn dynamic behaviors in order to generalize the obtained models.

Generally, the collected data for the detection and characterization process are noisy, but, the error tolerance ability of neural networks enables the detection scheme to differentiate the pattern from noise. This property is a huge advantage in fault detection and isolation problem resolution. In addition, similar patterns are separated using the property of characterization of a neural network.

3. Proposed Approach

Leak detection in intake systems is very difficult to achieve especially when the operating point corresponds to low load-torque couple. In these conditions, the compressor in the air path is not solicited by the driver, leading to similar pressure between the intake system and the atmosphere. This constraint requires an improved detection and characterization algorithm. The proposed system must increase the accuracy of the model, enhance the performances of the vehicle and guarantee the management of small leaks. In this paper, the Levenberg-Marquardt (LM) algorithm is proposed to carry out the detection and characterization tasks. The LM algorithm is used to train the air path diesel dynamics. Once the dynamics are modeled, the leak is detected and characterized by comparing the new measurements with the model established using a neural network. The proposed approach contains two blocks which are the training block and the decision block. This approach is shown in Figure 1.

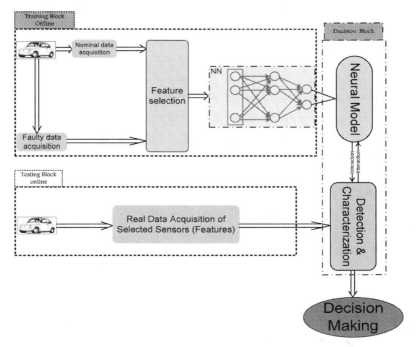

Figure 1. Detection and characterization scheme.

The proposed approach is designed to operate in on-line mode, thus, a classic process of real data acquisition is adopted in this work. It is important to remember that in this application we use only sensors selected by the feature selection algorithm and summarize the intake behavior of the diesel air path. The data acquired in this step are sent to the decision block in order to detect and characterize the leaks affecting the vehicle.

3.1. Training Block

3.1.1. Feature Selection

In this work, a very popular feature selection is chosen; it is the Chi-square algorithm [20]. Chi-square is a simple and general algorithm which achieves feature ranking using a discretization process. This algorithm is combined with a neural network classifier to select the features that we must keep.

The Chi-square algorithm is based on the X^2 which runs in two stages in this manner:

- **Stage 1:**
    1. Set the *sigLevel* to 0.5 for all features;
    2. Sort each feature according to its values;
    3. Compute the X^2 value for every pair of adjacent intervals, such that:
    4. Merge the pair of adjacent intervals with the lowest X^2 value until the X^2 value of each pair of adjacent intervals exceeds *sigLevel*.

$$X^2 = \sum_{i=1}^{2}\sum_{j=1}^{k} \frac{(A_{ij} - E_{ij})^2}{E_{ij}} \quad (1)$$

$$\text{with}: E_{ij} = R_i * C_j / N \quad (2)$$

where:

k: number of classes;

A_{ij}: number of samples in the ith interval and j^{th} class;

R_i: number of samples in the ith interval;

C_j: number of samples in the jth class;

N: total number of samples;

- **Stage 2:**
    1. Start with the *sigLevel0* corresponding to the last *sigLevel* value determined in the first stage;
    2. Associate *sigLevel(i)* with each feature and run merging;
    3. Consistency test:

If inconsistency < δ merge intervals and decrease *sigLevel* (*i*); Else, eliminate the ith features for the next step.

Firstly, the WEKA [21] data-mining tool is used to perform Chi-square ranking. The features are sorted according to their rank. Secondly, the most important features are selected using the neural network classifier, which will be described later. More precisely, the features will be eliminated iteratively from least important to most important and the weight of the eliminated features is evaluated according to the obtained classification Mean Squared Error (MSE).

3.1.2. Training

Pattern classification using neural networks consists of determining the class boundaries using the classifier. The training phase of neural networks achieves this goal. In this paper, a gradient-based training algorithm is used. This category of algorithm is most commonly used by researchers. One of these algorithms is Hessian-based algorithms; they can significantly reduce the convergence time. The Levenberg-Marquardt algorithm [22] belongs to category of Hessian-based techniques; it makes use of the advantages of Hessian-based algorithms in the optimization of nonlinear least squares.

The Levenberg-Marquardt algorithm is a well-known optimization technique. It locates the minimum of a function which is expressed by the sum of squares of nonlinear functions. This algorithm, widely used in many disciplines, is a combination of the Steepest-Descent and the Gauss-Newton methods. Depending on the distance between the current position and the best one, these techniques operate intermittently; if the current position is far from the best one the steepest-Descent is applied, otherwise the Gauss-Newton takes over. The Steepest-Descent technique used in the LM algorithm is slow, but it guarantees the convergence property. When the current position approaches the best one, the LM algorithm switches to the Gauss-Newton method which converges rapidly.

For neural network training, the objective function is the error of the type:

$$MSE = \frac{1}{2}\sum_{k=1}^{p}\sum_{i=1}^{n_0}(y_{kl} - a_{kl})^2 \qquad (3)$$

where y_{kl} are real data of diesel engines, a_{kl} are a network output, p is the total number of samples and n_0 represents the total number of nodes in the output layer.

In this work, the neural network used contains five layers. The first layer is the input layer which receives the data corresponding to the selected sensors which are used in this application. The three following layers are the hidden ones which represent the network core. The last one is the output layer which generates two signals when the detection task is considered or four signals when the tasks of detection and characterization are considered together.

The steps required in the neural network using the LM algorithm in batch-mode training are the following:

- Compute the corresponding network outputs and evaluate the mean square error for all inputs as in Equation (1);
- Calculate the Jacobian matrix $j(x)$, where x represents the weights and biases of the network;
- Solve the equation which adapts weights in order to obtain Δx, The update of the weighted vector Δx is computed as follows:

$$\Delta x = [J^T(x)J(x) + \mu I]^{-1}J^T(x)R \qquad (4)$$

where μ is the training parameter and R is a vector of size $pn0$ computed as follows:

$$R = \begin{pmatrix} y_{11} - a_{11} \\ y_{12} - a_{12} \\ \dots \\ y_{21} - a_{21} \\ y_{22} - a_{22} \\ \dots \\ y_{pn_0} - a_{pn_0} \end{pmatrix} \quad E_{ij} = R_i * C_j / N \quad (5)$$

$J^T(x)J(x)$ is referred to as the Hessian matrix.

- Recalculate the error using $x + \Delta x$. If there is the reduction of the error calculated in step 1, the training parameter µ is reduced by µ⁻, keep $x = x + \Delta x$ and return to step 1. If there is no reduction, increase µ by µ⁺ and go back to step 3. µ⁺and µ⁻ are fixed by the user;
- The algorithm is stopped in two cases; when the gradient is less than the predefined value, or when the error is below a given error objective.

Generally, the training step in neural networks is very complex and requires considerable computing resources, especially in on-line cases. In this work, the training problem is carried out in off-line mode, then, the obtained neural model is used to detect and characterize leakage in on-line mode. The adopted neural network returns both the nominal behavior corresponding to the system without leakage and the faulty system behavior (occurrence of leakage).

3.2. Decision Block

The decision block is the most essential component of the proposed scheme where the leaks in the intake of the air path are detected and characterized using the neural model developed in the training step. Direct interactions are established between the detection and characterization block with the neural network model in order to estimate the actual state of the system. The decision block works in two modes, "detection mode" or "detection and characterization mode". If the detection mode is considered, the decision block returns the two possible outputs, "No Leakage" or "Leakage". In "detection and characterization mode", it returns four outputs which are, "No Leak", "Low Leakage", "Medium Leakage" and "High Leakage".

4. APPLICATION

A critical operating mode system is considered to illustrate the effectiveness of the proposed approach. This mode concerns the case of low engine load, speed and couple, where the leak detection and characterization problem is not systematically realized. In this application, the data acquisition is carried out using xMOD software.

4.1. xMOD Software

xMOD is a software platform that was developed at "IFP Energies Nouvelles" combining heterogeneous models and a virtual experimentation laboratory. These heterogeneous models are generated by different simulation tools, such as Matlab/Simulink, AMESim, Dymola, SimulationX and GT Power. A combination of these tools means benefiting from the advantages of each modeling and simulation tool, and the user can freely select these tools.

In this work, xMOD is used to simulate the diesel engine functioning, especially the air path behavior. The simulation model produced by IFP "Energies Nouvelles" is used, to which a leak model has been added. The diameter of the leak can be freely adjusted. The simulation results can be recovered and stored in text files.

4.2. MSE: All Features vs. Selected Features

Before presenting the results with selected features, a comparison of the MSE evaluation for "all features" and "selected features" is presented in Table 1.

In order to illustrate the advantage of feature selection, the MSE values are jointly shown with their training run times. In this table, we can firstly observe that MSE values corresponding to the use of all features are greater than the MSE values when only the selected features are used. Secondly, we observe that the run time corresponding to the use of all features is always higher than when the selected features are used. For example, when the torque value is set to 40 Nm, all MSE values of the all features case are greater than those corresponding to the selected features case. The same conclusion can be drawn for the remaining three cases except for some values. The detection and characterization tasks are presented for the selected features case.

Table 1. Mean Squared Error (MSE) Evolution with torque variation.

Torque	40 Nm (MSE/Run Time)		110 Nm (MSE/Run Time)		130 Nm (MSE/Run Time)		150 Nm (MSE/Run Time)	
Leaks	All Features	Selected Features	All Features	Selected Features	All Features	Selected Features	All Features	Selected Features
0.1mm	0.208/14'59"	0.0221/10'35"	0.168/12'40"	0.0166/10'41"	0.167/14'50"	0.0109/8'40"	0.00547/15'06"	0.00799/8'59"
0.2mm	0.0449/16'29"	0.0189/8'49"	0.00790/12'44	0.0147/8'55"	0.00988/13'10"	0.0163/8'45"	0.00083/15'47"	0.000812/10'15"
0.3mm	0.0270/18'44"	0.0114/9'19"	0.00608/12'52"	0.00871/9'44"	0.00298/16'23"	0.00266/8'45"	0.00430/14'17"	0.00165/10'03"
0.4mm	0.0205/15'24"	0.0204/9'25"	0.00239/14'39"	0.00413/10'15"	0.00219/13'55"	0.00401/10'53"	0.00273/13'23"	0.00419/11'38"
0.5mm	0.0186/18'41"	0.0127/8'34"	0.00458/14'02"	0.00369/7'57"	0.00156/12'22"	0.00222/10'23"	0.00208/13'44"	0.00428/11'25"
0.6mm	0.0121/15'50"	0.00807/9'30"	0.00447/14'17"	0.00315/9'52"	0.00351/14'12"	0.00104/9'34"	0.00656/15'00"	0.00221/11'04"
0.7mm	0.0180/15'57"	0.00404/9'43"	0.00669/16'11"	0.00234/9'07"	0.00100/8'02"	0.00102/8'40"	0.00601/14'33"	0.000873/10'01"
0.8mm	0.00867/17'26"	0.00727/8'21"	0.00584/13'42"	0.00198/9'54"	0.00099/7'18"	0.00101/9'56"	0.000897/13'35"	0.00193/9'52"
0.9mm	0.00618/14'30"	0.00910/8'56"	0.00451/13'32"	0.00415/9'33"	0.00211/13'28"	0.00131/8'58"	0.00270/13'36"	0.00087/11'29"
1.0mm	0.00989/13'11"	0.00638/9'14"	0.00217/15'01"	0.00111/9'43"	0.00102/13'10"	0.00099/5'09"	0.00392/13'15"	0.000896/8'41"

4.3. Detection Task Results

The first property of the proposed approach is the detection ability. In this situation, the neural network trains two classes: "No Leakage mode" and "Leakage mode". The training set consists of 10,000 samples without leaks and 10,000 samples with leaks. We choose three values of leaks, 0.1 mm, 0.4 mm and 0.9 mm. These results are obtained using only selected features in the training algorithm.

5. INTERPRETATION

Figure 2, Figure 3, Figure 4, Figure 5, Figure 6, Figure 7, Figure 8, Figure 9, Figure 10 show the effectiveness of the proposed approach where we can see that the leak is detected for all considered diameters. Mean Squared Error (MSE) values give information about the accuracy of the neural network used. From the obtained results we can first remark that the MSE values increase when the torque values decrease. For example, in the first case (Figure 2, Figure 3, Figure 4) when the leak diameter is set to 0.1 mm, the MSE value decreases from 0.0240 (2.4%) to 0.00746 (0.7%) when the torque increases from 110 Nm to 150 Nm. This observation can be explained by the fact that the air path system (compressor) works at a lower speed. In other words, in low speed, the mechanical compressor of the air path is not solicited. The same remark is applied to cases 2 and 3.

Naturally, the leak is easily detected when it is large, but, it becomes extremely difficult to detect when it is very small. The obtained results show that the proposed approach is efficient and the leak is detected in all cases even when it is equal to 0.1 mm (almost negligible leakage). In addition, this approach gives better results at higher engine load, speed and couple which is expected due to higher flow on the air intake system.

Case1: Leak = 0.1 mm

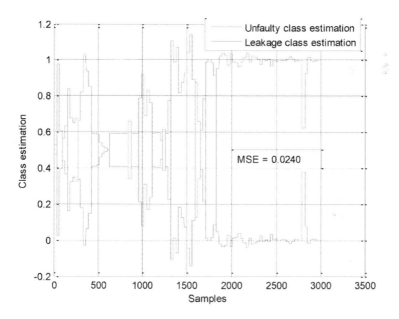

Figure 2. Engine_speed = 1000 rpm and torque = 110 Nm.

5. Interpretation

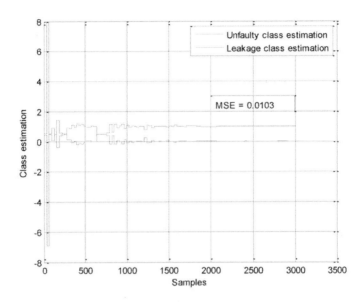

Figure 3. Engine_speed = 1000 rpm and torque = 130 Nm.

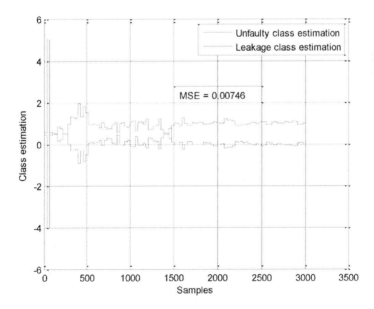

Figure 4. Engine_speed = 1000 rpm and torque = 150 Nm.

Case2: Leak = 0.4 mm

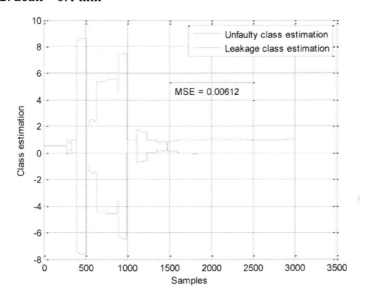

Figure 5. Engine_speed = 1000 rpm and torque = 110 Nm.

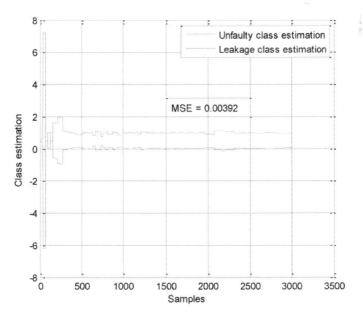

Figure 6. Engine_speed = 1000 rpm and torque = 130 Nm.

5. Interpretation

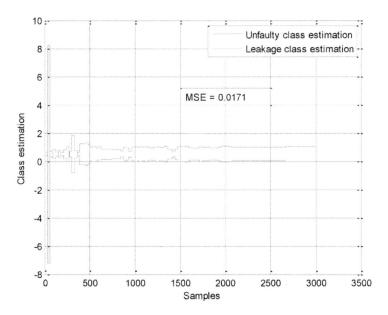

Figure 7. Engine_speed = 1000 rpm and torque = 150 Nm.

Case3: Leak = 0.9 mm

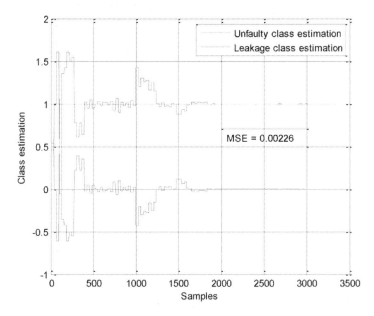

Figure 8. Engine_speed = 1000 rpm and torque = 110 Nm.

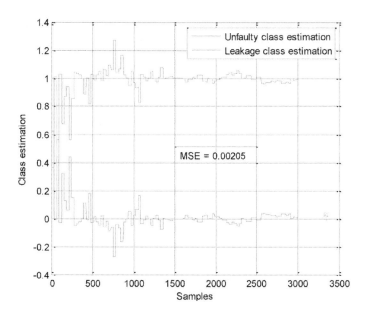

Figure 9. Engine_speed = 1000 rpm and torque = 130 Nm.

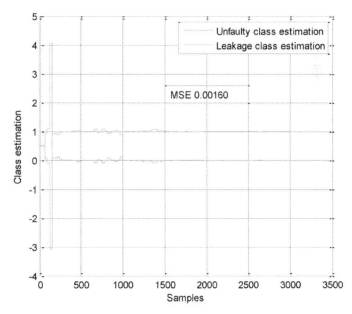

Figure 10. Engine_speed = 1000 rpm and torque = 150 Nm.

Characterization Task Results

This property is very important and allows for the estimation of the severity of the leak. Thus, the characterization of the leak diameter is highly desired and it is often difficult to accomplish. In order to do this, the neural network is trained with the data of four modes which are: "No Leak", "Low Leakage", "Medium Leakage" and "High Leakage". The three last modes respectively correspond to "Leak = 0.1 mm", "Leak = 0.4 mm" and "Leak = 0.9 mm". The data of each mode contains 10,000 samples. The Figure 11, Figure 12, Figure 13 show the leak characterization in each mode.

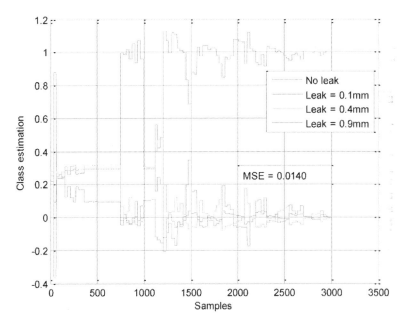

Figure 11. Leak Characterization in "Low Leakage" case.

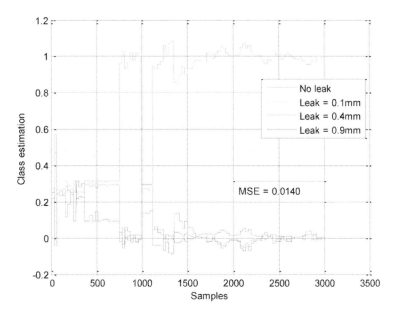

Figure 12. Leak Characterization in "Medium Leakage" case.

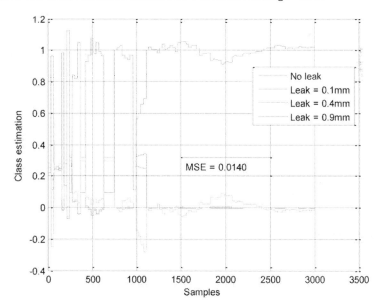

Figure 13. Leak Characterization in "High Leakage" case.

These figures show the effectiveness of the proposed approach dealing with leak characterization. Firstly, we remark that MSE value (0.0140) is low and the accuracy of the neural network is demonstrated. Secondly, the leakage class is found for each case.

6. CONCLUSIONS

A leak detection and characterization approach for diesel air paths has been developed. The proposed approach contains two blocks: a training block and a decision block. The first one is realized off-line and combines a feature selection algorithm with a neural network based on the Levenberg-Marquardt optimization. The L-M function was chosen for its accuracy and adaptability; it combines two different techniques according to the current position of the solution compared to the best one. The second block uses the neural model obtained in training phase in order to detect and characterize leaks that appear in the air path system. The detection and characterization capability is evaluated using the MSE index.

The proposed approach effectively solves the leak detection and characterization problem, especially in the case of small leaks in critical operating points (low speed and torque). In order to validate this solution, the proposed algorithms will be implemented in a real diesel engine.

ACKNOWLEDGMENTS

The authors gratefully thank the ANR (Agence Nationale de la Recherche), "Haute Normandie" Region and FEDER for financially supporting this work.

AUTHOR CONTRIBUTIONS

This paper proposes a model-free diagnosis approach for leak detection and characterization in Diesel Engines. The main contribution in this work is the capacity of the proposed approach to detect a very small leak in the air path equal to 0.1 mm in diameter. The minimum leak detected with other approaches is equal to 4 mm in diameter [23]. The proposed approach also gives good results for leak characterization in static mode.

Other investigations are in progress for the characterization task in dynamical cases and for sensor fault diagnosis.

Ghaleb Hoblos was the scientific responsible of this work within the ANR project DIVAS "**D**iesel **I**nnovative **VV**A and**A**dvanced air **S**ystem for Downspeeding". Ghaleb Hoblos and Mourad Benkaci conceived and designed the proposed approach. Mourad Benkaci performed most of the tests. Ghaleb Hoblos and Mourad Benkaci wrote the paper.

REFERENCES

1. Nyberg, M.; Nielsen, L. Model-based diagnosis for the air intake system of the SI-engine. In Proceedings of the SAE International Congress and Exposition, Detroit, MI, USA, 24–27 February 1997.
2. Nyberg, M. Model-based diagnosis of an automotive engine using several types of fault models. *IEEE Trans. Control Syst. Technol.* **2002**, *10*, 679–689.
3. Nyberg, M. Model-based diagnosis of the air path of an automotive diesel engine. *Control Eng. Pract.* **2004**, *12*, 513–525.
4. Gertler, J.; Costin, M.; Fang, X.; Kowalczuk, Z.; Kunwer, M.; Monajemy, R. Model-based diagnosis for automotive engine—Algorithm development and testing on a production vehicle. *IEEE Trans. Control Syst. Technol.* **1995**, *13*, 61–69.
5. Isermann, R. Process fault detection based on modeling and estimation methods: A survey. *Automatica* **1984**, *20*, 387–404.
6. Sorsa, T.; Koivo, H.N. Application of artificial neural networks in process fault diagnosis. *Automatica* **1993**, *29*, 843–849.
7. Capriglione, D.; Liguori, C.; Pianese, C.; Pietrosanto, A. On line sensor fault detection, isolation and accommodation in automotive engines. *IEEE Trans. Instrum. Meas.* **2003**, *52*, 1182–1189.
8. Sangha, M.S.; Yu, D.L.; Gomm, J.B. On-Board monitoring and diagnosis for spark ignition air path via adaptive neural networks. *J. Automob. Eng.* **2006**, *220*, 1641–1655.
9. Chen, P.C. A novel diagnostic system for gasoline-engine leakage detection. *J. Automob. Eng.* **2011**, *225*, 225–685.
10. Christina, B.V; Tshilidzi, M. Application of Feature Selection and Fuzzy ARTMAP to Intrusion Detection. In Proceedings of the

IEEE International Conference on Systems, Man and Cybernetics, Taipei, Taiwan, 8–11 October 2006.

11. Carpenter, G.A.; Grossberg, S.; Markuzon, N.; Reynolds, J.H.; Rosen, D.B. Fuzzv ARTMAP: A Neural Network Architecture for Incremental Supervised Learning of Analog Multidimentional Maps. *IEEE Trans. Neural Netw.* **1992**,*3*, 698–713.

12. Wang, L.; Yu, J. Fault Feature Selection Based on Modified Binary PSO with Mutation and its Application in Chemical Process Fault diagnosis. In Proceedings of the International Conference on Advanced in Natural Computation N1, Changsha, China, 27–29 August 2005; pp. 832–840.

13. Xu, Z.; Xuan, J.; Shi, T.; Hu, Y. Application of a modified fuzzy ARTMAP with feature-weight learrning for the fault diagnosis of bearing. *Expert Syst. Appl.* **2009**, *36*, 9961–9968.

14. Sugumaran, V.; Muralidharan, V.; Ramachandran, K.I. Feature selection using Decision Tree and classification through Proximal Support Vector Machine for fault diagnostics of roller bearing. *Mech. Syst. Signal Process.* **2007**, *21*, 930–942.

15. Casimir, R.; Boutleux, E.; Clerc, G.; Yahoui, A. The use of features selection and nearest neighbors rule for faults diagnostic in induction motors. *Eng. Appl. Artif. Intell.* **2006**, *19*, 169–177.

16. Chiang, L.H.; Pell, R.J. Genetic algorithms combined with discriminant analysis for key variable identification. *J. Process Control* **2004**, *14*, 143–155.

17. Verron, S.; Tiplica, T.; Kobi, A. Fault detection and identification with a new feature selection based on mutual information. *J. Process Control* **2008**, *18*, 479–490.

18. Mao, Y.; Xia, Z.; Yin, Z.; Sun, Y.; Wan, Z. Fault Diagnosis Based on Fuzzy Support Vector Machine with Parameter Tuning and Feature Selection. *Chin. J. Chem. Eng.* **2007**, *15*, 233–239.

19. Wang, K. Neural Network Approach to Vibration Feature Selection and Multiple Fault Detection for Mechanical Systems. In Proceedings of the ICICIC'06 First International Conference on Innovative Computing, Information and Control, Beijing, China, 30 August–1 September 2006; Volume 3, pp. 233–239.

20. Alexandrov, M.; Gelbukh, A.; Lozovo, G. Chi-square Classifier for Document Categorization. In Proceedings of the 2nd International Conference on Intelligent Text Processing and

Computational Linguistics, Mexico City, Mexico, 18–24 February 2001.

21. Witten, I.H.; Eibe, F. *Data Mining: Practical Machine Learning Tools and Techniques*, 2nd ed.; Morgan Kaufmann Publishers Inc.: San Francisco, CA, USA, 2005.

22. Levenberg, K. A Method for the Solution of Certain Non-linear Problems in Least Squares. *Q. Appl. Math.* **1944**, *2*, 164–168.

23. Ceccarelli, R.; Canudas-de-Wit, C.; Moulin, P.; Sciarretta, A. Model-based Adaptive Observers for Intake Leakage Detection in Diesel Engines. In Proceedings of the IEEE ACC'09 American Control Conference, Saint Louis, MO, USA, 10–12 June 2009.

CHAPTER 4

Homogeneous Charge Compression Ignition Combustion: Challenges and Proposed Solutions

Mohammad Izadi Najafabadi and Nuraini Abdul Aziz

Department of Mechanical and Manufacturing Engineering, Faculty of Engineering, Universiti Putra Malaysia (UPM), 43400 Serdang, Selangor, Malaysia

ABSTRACT

Engine and car manufacturers are experiencing the demand concerning fuel efficiency and low emissions from both consumers and governments. Homogeneous charge compression ignition (HCCI) is an alternative combustion technology that is cleaner and more efficient than the other types of combustion. Although the thermal efficiency and NO_x emission of HCCI engine are greater in comparison with traditional engines, HCCI combustion has several main difficulties such as controlling of ignition timing, limited power output, and weak cold-start capability. In this study a literature review on HCCI engine has been performed and HCCI challenges and proposed solutions have been investigated from the point view of Ignition Timing that is the main problem of this engine. HCCI challenges are investigated by many IC engine researchers during the last decade, but practical solutions have not been presented for a fully HCCI engine. Some of the solutions are slow response time and some of them are technically difficult to implement. So it seems that fully HCCI engine needs more investigation to meet its mass-production and the future research and application should be considered as part of an effort to achieve low-temperature combustion in a wide range of operating conditions in an IC engine.

1. INTRODUCTION

Although electric and hybrid vehicles (EVs and PHEVs) have emerged on the market, still the internal combustion engines are the most popular automotive power plant. However, in recent decades, serious concerns have piled up considering the environmental impact of the gaseous and particulate emissions arising from operation of these engines. As a result, ever tightening legislation, that restricts the levels of pollutants that may be emitted from vehicles, has been introduced by governments around the world. In addition, concerns about the world's finite oil reserves and CO_2 emissions have led to heavy taxation of road transport, mainly via on duty on fuel. These factors have led to massive pressure on vehicle manufacturers to research, develop, and produce ever cleaner and more fuel-efficient vehicles [1].

Over the last decade, an alternative combustion technology, commonly known as homogeneous charge compression ignition (HCCI), has emerged and it has the potential to decrease emissions and fuel consumption in transportation [2, 3]. HCCI is a clean and high efficiency technology for combustion engines that can be scaled to any size-class of transportation engines as well as used for stationary applications [4]. These benefits of HCCI (especially relative to spark ignition engines) are acquired by virtue of lean/dilute operation.

The two dominating engine concepts commonly used today are the diesel and SI engines. A comparison between the two engines shows that the SI engine equipped with a catalytic converter provides low emissions but lacks in efficiency. The diesel engine on the other hand provides high efficiency but also produces high emissions of NO_x and particles. An engine concept capable of combining the efficiency of a diesel engine with the tailpipe emissions level of an SI engine is the homogeneous charge compression ignition (HCCI) engine [5]. In other words, HCCI is the autoignition of a homogeneous mixture by compression.

The following literature review has focused on HCCI challenges and proposed solutions from the point view of Ignition Timing as the most critical problem of HCCI engine. This point of view has been tried to be discussed through the paper as its particular characteristic. At first, previous studies in the field of HCCI engine including two-stroke and four-stroke HCCI engines are discussed. Next, HCCI challenges and proposed solutions are reviewed. Finally, HCCI ignition timing as the most important problem of HCCI is considered and the main controlling methods such as mixture dilution, changing fuel properties, fast thermal management, and direct injection are presented.

2. HCCI/CAI ENGINE

The homogeneous charge compression ignition (HCCI) or controlled autoignition (CAI) combustion has often been considered a new combustion process amongst the numerous research papers published over the last decade. However, it has been around perhaps as long as the spark ignition (SI) combustion in gasoline engine and compression ignition (CI) combustion in diesel engines [1].

In the case of gasoline engines, the HCCI combustion had been observed and was found responsible for the "after-run"/"run-on" phenomenon that many drivers had experienced with their carbureted gasoline engines in the sixties and seventies, when a spark ignition engine continued to run after the ignition was turned off [1].

In the case of diesel engines, the hot-bulb oil engines were invented and developed over 100 years ago. In these engines, the raw oil was injected onto the surface of a heated chamber called hot-bulb. This early injection gives the fuel lots of time to vaporize and mix with air. The hot-bulb had to be heated on the outside for the start-up and once the engine had started, the hot-bulb was kept hot by using the burned gases. Later design placed injection through the connecting passage between the hot-bulb and the main chamber so that a more homogeneous mixture could be formed, resulting in auto-ignited homogeneous charge combustion [6].

2.1. Two-Stroke HCCI Engine

For solving one of the main problems of the two-stroke engine which was the unstable, irregular, and incomplete part load combustion responsible for excessive emissions of unburned hydrocarbons, a significant research work was performed from the end of the 1960s to the end of the 1970s [1]. Lots of studies were performed during this period by Jo et al. to investigate the part load lean two-stroke combustion [7]. He found that the irregularities of the combustion and the autoignition which were considered as the weak points of the two-stroke engine could be effectively controlled. This period was successfully concluded by the innovative work he published with his colleague, Onishi et al. who managed to get a part load stable two-stroke combustion process for lean mixtures in which ignition occurs without spark assistance [8]. Remarkable improvements in stability, fuel efficiency, exhaust emissions, noise, and vibration were reported. Onishi and his colleagues called this new combustion process "ATAC" (Active Thermo-Atmosphere Combustion). The first electric generator using an ATAC two-stroke engine was then commercialized in Japan from this period during a few years as shown in Figure 1.

2. HCCI/CAI Engine

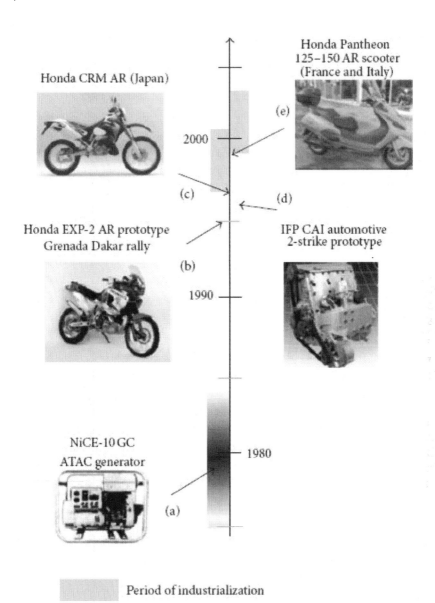

Figure 1. History of production and most advanced prototype CAI two-stroke engines [1].

Another paper concerning two-stroke autoignition was published in 1979 [9]. Noguchi and his colleagues named this autoignition combustion the TS (Toyota-Soken) combustion process. They also concluded that TS combustion occurred similarly without flame front while showing great efficiency and low emissions. They were one of the first to suggest that active radicals in residual gases could play an important role in the autoignition process.

In the late 1980s, Duret tried to apply Onishi's pioneering work to DI two-stroke engines for improvement of part load emissions. For this purpose, he investigated the idea of using a butterfly exhaust throttling valve as previously shown by Tsuchiya et al. in a carburetted engine [10]. The first application of ATAC autoignition with direct fuel injection engine was then described in 1990 [11]. CFD calculations showed that mixing between the residual gas and fresh intake air may be reduced by precisely regulating the introduction of the intake flow through the use of an exhaust control valve [1].

This research work was further developed until the mid-1990s and the interest of using transfer port throttling (the transfer duct in a two-stroke engine is the duct in which the fresh charge is transferred from the pump crankcase to the combustion chamber through a port on the wall of the cylinder) to even better control the degree of mixing between the fresh charge and the hot and reactive residual gas was demonstrated [1].

As shown in Figure 1, the first automotive two-stroke direct injection engine prototype using the transfer port throttling technique (a transfer duct for better controlling the degree of mixing between the fresh charge and the hot and reactive residual gas) for running in controlled autoignition (CAI) was presented by Duret and Venturi in 1996 [12]. Considering the benefits of combining direct injection with CAI, this engine was easily able to meet the European emissions standards valid up to the year 2000 without NO_x after treatment and with more than 20% fuel economy improvement compared to its four-stroke counterpart of equivalent power output [1].

In this period the possibility of using the autoignition in two-stroke motorcycle engines was investigated by Ishibashi. He showed that by using a charge control exhaust valve it was possible to control the amount of active residual gases in the combustion chamber as well as in cylinder pressure before compression [13]. He called this combustion process "Activated Radicals combustion (AR combustion)." Honda EXP-2 400 cc AR prototype was prepared for the 1995 Grenada-Dakar rally and performed very well compared to the four-stroke motorcycles, thanks in particular to their high fuel economy. This work was further developed [14, 15] up to the first industrial application of AR combustion in production in a Japanese motorcycle model in 1996 and in a European scooter model in 1998 (Figure 1) [1].

Recently, in 2008, Ricardo has developed a new prototype engine called 2/4 SIGHT which uses HCCI concept. This gasoline engine concept uses novel combustion, boosting, control, and valve actuation technologies to enable automatic and seamless switching between two- and four-stroke operations, with the aim of delivering significant performance and fuel economy improvements through aggressive downsizing. An engine equipped with this new system is capable of running on either the 2- or 4-stroke engine cycle, allowing their V6 test-bed to be downsized from 3.5 liters to 2.0 liters while making the same power output. This downsizing leads to a 27% reduction in fuel consumption and correspondingly lowered emissions. This engine is shown in Figure 2.

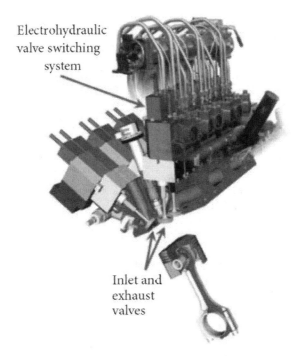

Figure 2. Ricardo 2/4 sight engine [16].

A further recent HCCI engine was reported by Lotus in 2008 [16]. As shown in Figure 3, a single-cylinder research engine called OMNIVORE has been built, employing loop scavenging and direct injection with the ability to vary geometrically the compression ratio from $8:1$ to $40:1$ or from $6.4:1$ to $24.4:1$ on a trapped basis (after exhaust port closure).

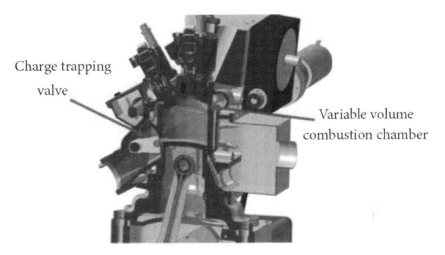

Figure 3. Lotus OMNIVORE two-stroke engine [16].

Blundell et al. and Turner et al. have published this engine data showing very low NO_x emission levels and a minimum part-load indicated a specific fuel consumption of 218 g/kW h using gasoline and 217 g/kW h using E85 [17, 18]. The engine was designed to be able to operate in HCCI modes and is intended to explore CO_2 reduction and the ability to operate on alternative alcohol-based fuels and gasoline, allowing flexible fuel vehicle operation.

2.2. Four-Stroke HCCI Engine

Based on the previous work in two-stroke engines [8], in 1983 Najt and Foster extended the work to four-stroke engines and attempted to gain additional understanding of the underlying physics of HCCI combustion [19]. They are the first to apply HCCI combustion concept in a four-stroke gasoline engine. In this work they considered that HCCI is controlled by chemical kinetics, with negligible influence of turbulence and mixing. They conducted experiments using PRF fuels and intake preheating. By means of heat release analysis and cycle simulation, they pointed out that HCCI combustion process was governed by low temperature (smaller than 950°K) hydrocarbon oxidation kinetics. Also they concluded that HCCI combustion is a chemical kinetic combustion process controlled by the temperature, pressure, and composition of the in-cylinder charge.

In 1989, Thring further extended the work of Najt and Foster in four-stroke engines by examining the performance of an HCCI engine operated with a full-blended gasoline [20]. The operating regime of a single-cylinder engine was mapped out as a function of air fuel equivalence ratio, EGR rate, and compression ratio.

Studies on four-stroke engines have shown that it is possible to achieve high efficiencies and low NO_x emissions by using a high compression ratio and lean mixtures [21]. In the four-stroke case, a number of experiments have been performed where the HCCI combustion in itself is studied. This has mostly been done with single cylinder engines, which normally do not provide brake values. However, Stockinger demonstrated brake efficiency of 35% on a 4-cylinder 1.6 liter engine at 5 bar Brake Mean Effective Pressure (BMEP) [22]. Later studies have shown brake thermal efficiencies above 40% at 6 bar BMEP [23].

3. HCCI/CAI CHALLENGES AND PROPOSED SOLUTIONS

Although advantageous over traditional engines in thermal efficiency and NOx emission, HCCI combustion has several main difficulties. These difficulties include "control of combustion timing," "limited power output," "homogenous mixture preparation," "high unburned Hydrocarbon (HC) and carbon monoxide (CO) emissions," and "weak cold-start capability" [4].

HC and CO emissions of HCCI engine are relatively higher in comparison with those of diesel engines [24]. Some potential exists to mitigate these emissions at high load by using direct in-cylinder fuel injection to achieve appropriate partial-charge stratification. However, in most cases, controlling HC and CO emissions from HCCI engines will require exhaust emission control devices where fuel optimization was not used. Catalyst technology for HC and CO removal is well understood and has been standard equipment on automobiles for many years. However, the cooler exhaust temperatures of HCCI engines may increase catalyst light-off time and decrease average effectiveness. As a result, meeting future emission standards for HC and CO will likely require further development of oxidation catalysts for low-temperature exhaust steams. However, HC and CO emission control devices are simpler, more durable, and less dependent on scarce, expensive precious metals than are NO_x and PM emission control devices [25]. Thus, simultaneous chemical oxidation of HC and CO in an HCCI engine is much easier than simultaneous chemical reduction of NO_x and oxidation of PM in a Compression-Ignition Direct-Injection (CIDI) engine.

At cold start, the compressed-gas temperature in an HCCI engine will be reduced because the charge receives no preheating from intake manifold and the compressed charge is rapidly cooled by heat transferred to the cold combustion chamber walls. Without some compensating mechanism, the low compressed-charge temperatures could prevent an HCCI engine from firing. Various mechanisms for cold-starting in HCCI mode have been proposed, such as using glow plugs, using a different fuel or fuel additive, and increasing the compression ratio using variable compression ratio (VCR) or variable valve timing (VVT). Perhaps the practical approach would be to use Spark Assisted Compression Ignition (SACI) approach as a bridge to the gap between HCCI and SI engines [26]. For engines equipped with VVT, it may be possible to make this warm-up period as short as a few fired cycles, since high levels of hot residual gases could be retained from previous spark ignited cycles to induce HCCI combustion. Although solutions appear feasible, significant research and developing will be required to advance these concepts and prepare them for production engines [27].

Table 1 lists three major HCCI challenges and solutions proposed to address specific problems. The problem of high HC and CO emissions in HCCI is also linked to control of combustion timing since HC and CO emissions highly depend on the location of ignition timing. Despite the plurality of different proposed solutions, each of the proposed solutions has its own drawbacks. Variable intake temperature, variable intake pressure, and variable coolant temperature have slow response time, while VCR and VVT are technically difficult to implement. Practicality and cost effectiveness are the main concerns with most of the proposed options such as water injection and modulating two or more fuels [4].

As mentioned (Table 1), the main problem of HCCI is control of HCCI combustion timing. To have more discussion, this problem and its proposed solutions are the subject of the next part of this study.

4. CONTROL OF HCCI IGNITION TIMING

Several strategies have been investigated, with various levels of success, for controlling HCCI combustion timing and extending the load range. Most of these strategies can be divided into the broad categories of mixture dilution, modifying fuel properties, fast thermal management, and in-cylinder direct fuel injection. Many studies investigating HCCI control employ more than one method due to the complicated and highly coupled nature of the HCCI combustion problem [71].

Table 1. Main HCCI challenges and proposed solutions [4].

HCCI challenges	Proposed solutions

Control of combustion timing	(i) Changing temperature history of mixture:
	  (a) VVT and residual/exhaust gas trapping
	    (1) Exhaust gas trapping [28, 29]
	    (2) Modulating intake and exhaust flows [30, 31]
	    (3) Combination of both [32]
	  (b) Variable compression ratio (VCR) [33–36]
	  (c) Variable EGR [31, 37, 38]
	  (d) In-cylinder injection timing [39–41]
	  (e) Modulating intake temperature [42–44]
	  (f) Water injection [45]
	  (g) Variable coolant temperature [46]
	(ii) Changing mixture reactivity:
	  (a) modulating two or more fuels [21, 47–49]
	  (b) fuel stratification [50–54]
	  (c) fuel additives and reforming [55–57]
	  (d) variable EGR [37, 38, 58]

Limited power output	(i) Boosting intake air flow:
	  (a) supercharging [35, 59–61]
	  (b) turbocharging [61–63]
	(ii) Dual-mode engines (HCCI at low load):
	  (a) SI-HCCI [58, 64, 65]
	  (b) diesel-HCCI [66, 67]

Homogenous mixture preparation	(i) Fuel injection in a highly turbulent port flow for gaseous and highly volatile fuels [68, 69] (ii) Early in-cylinder injection with sophisticated fuel injectors for diesel fuels [60, 70]

4.1. Mixture Dilution for HCCI Control

In order to achieve CAI/HCCI combustion, high intake charge temperatures and a significant amount of charge dilution must be present. In-cylinder gas temperature must be sufficiently high to initiate and sustain the chemical reactions leading to autoignition processes. Substantial charge dilution is necessary to control runaway rates of the heat releasing reactions. Both of these requirements can be realized by recycling the burnt gases within the cylinder.

One approach to HCCI combustion phasing control is to advance or retard combustion timing by diluting the cylinder mixture. Najt and Foster showed that HCCI combustion in a four-stroke engine could be controlled by introducing re-circulated exhaust gas into the cylinder intake mixture [19]. Christensen and Johansson showed combustion timing to be slower with higher amounts of EGR [72].

The presence of the recycled gases has a number of effects on the CAI combustion and emission processes within the cylinder. Firstly, if hot burnt gases are mixed with cooler inlet mixture of fuel and air, the temperature of the intake charge increases owing to the heating effect of the hot burnt gases. This is often the case for CAI combustion with high octane fuels, such as gasoline and alcohols. Secondly, the introduction or retention of burnt gases in the cylinder replaces some of the inlet air and hence reduces the oxygen concentration (specially with a large amount of EGR). The reduction of air/oxygen due to the presence of burnt gases is called the dilution effect. Thirdly, the total heat capacity of the in-cylinder charge will be higher with burnt gases, mainly owing to the higher specific heat capacity values of carbon dioxide (CO_2) and water vapor (H_2O). This rise in the heat capacity of the cylinder charge is responsible for the heat capacity effect of the burnt gases. Finally, combustion products present in the burnt gases can participate in the chemical reactions leading to autoignition and subsequent combustion. This potential effect is classified as the chemical effect [1].

EGR or recycling of burned gases is the most effective way to moderate the pressure rise rate and expand the HCCI operation to higher load regions. The studies done related to EGR include both external EGR and internal EGR (residual combustion products) to achieve proper combustion phasing. External EGR is the more commonly utilized method for recycling exhaust gases. However, external EGR control has issues, such as, slow response time and difficulties in handling transient operating conditions [73]. A second way of reintroducing exhaust gases is through internal exhaust gas recirculation where the amount of exhaust gas residual in the cylinder is varied by changing the timing of the intake and exhaust valve's opening and closing events.

4.1.1. External Exhaust Gas Recirculation

External exhaust gas recirculation has been investigated by many researchers in the last decades. The study done by Thring investigated the effects of EGR rate (between 13 and 33%) on the achievable HCCI operating range and engine-out emissions [20]. Their study found out that the maximum load of HCCI operating range for a four-stroke engine was less than that of a two-stroke engine under the selected conditions.

Christensen and Johansson observed that the upper load limit of a supercharged HCCI engine could be increased to an IMEP of 16 bars through the addition of approximately 50% EGR to the intake mixture, which retarded combustion and avoided knock [74]. In this study high EGR rates were used in order to reduce the combustion rate. While external EGR is promising for load range and combustion phasing improvement, some drawbacks still exist. For recirculation of the exhaust gas into the intake mixture, the exhaust manifold pressure has to be increased to a level over that of the intake manifold pressure. This pressure increase is often achieved by throttling the exhaust manifold, which can result in higher pumping losses and thus an overall lower net efficiency of the engine. Efficiency losses are also seen as a result of cooling the exhaust gases before reinduction to prevent early autoignition [74].

In 2001, Morimoto et al. found similar results using a Natural Gas fueled engine [75]. In this study external cooled EGR was used to control combustion phasing and extend the load range of an HCCI engine. He also concluded that the total hydrocarbon emissions were reduced at higher loads with the introduction of EGR.

Numerical studies conducted by Narayanaswamy and Rutland, using a multizone model coupled with GT-Power, confirmed that the effects of EGR (external) on diesel HCCI operation vary with different levels of EGR [76]. Interestingly they pointed out that ignition was advanced initially for low EGR cases and then began to retard with increase in EGR percentage. The effect of cold EGR on the start of combustion was explained by competing effects, with the increase of the equivalence ratio advancing the ignition timing and the diluting effects retarding the combustion. As the EGR increases, the advancing effect prevails at first, and then evidently the retarding effect becomes dominant for further increase in EGR.

Atkins and Koch also observed that diluting the intake mixture using EGR is effective in retarding SOC timing. Similarly the introduction of EGR (around 62%) resulted in increasing maximum gross efficiency to 51%, much higher than that which could be achieved in an SI engine [77].

In 2011, Fathi et al. investigated the influence of external EGR on combustion and emissions of HCCI engine [38]. In his study, a Waukesha Cooperative Fuel Research (CFR) single cylinder research engine was used to be operated in HCCI combustion mode fueled by natural gas and n-Heptane. The main goal of the experiments was to investigate the possibility of controlling combustion phasing and combustion duration using various Exhaust Gas Recirculation (EGR) fractions. The influence of EGR on emissions was discussed. Results indicated that applying EGR reduces mean charge temperature and has profound effect on combustion phasing, leading to a retarded Start of Combustion (SOC) and prolonged burn duration. Heat transfer rate decreases with EGR addition. Under examined condition EGR addition improved fuel economy, reduced NOx emissions, and increaseo HC and CO emissions.

4.1.2. Internal Exhaust Gas Recirculation

Internal exhaust gas recirculation is another promising method for achieving stable HCCI combustion. By changing the valve timing of the engine the amount of trapped residual gases (TRG) in the cylinder can be changed, thereby changing the temperature, pressure, and composition of the cylinder mixture at IVC. In 2001, Law et al. found that it was possible to change the amount of internal EGR by varying valve timing, which in turn allows for control of combustion phasing of HCCI combustion [28].

A systematic study on the effects of internal EGR was carried by Zhao et al. in a four-stroke gasoline HCCI engine via analytical and experimental approaches [78]. He revealed that the charge heating effect of the hot recycled gases was mainly responsible for advancing the autoignition timing and reducing the combustion duration. The dilution effect extended the combustion duration but had no effect on the ignition timing. The total heat capacity of the in-cylinder charge with EGR (internal) was found to rise due to the presence of species with higher specific heat capacity, such as CO_2 and H_2O. This effect reduced the heat release rate, thereby increasing the combustion duration. Furthermore, the EGR chemical effect was shown to have no influence on the autoignition timing and heat release rate but slightly reduced the combustion duration at high concentration of burned gases.

Milovanovic et al. studied the influence of a fully variable valve timing (VVT) strategy on the control of a gasoline HCCI engine and found that EVC and IVO timing have the greatest impact on the ability to control HCCI combustion timing [79]. EVO and IVC timing were found to have little effect on HCCI combustion phasing control. A different research on fully VVT control of HCCI combustion was seen in the research of Urata et al. where a combination of direct injection, fully VVT with an electromagnetic valve train, and intake boost was used to control HCCI [80]. He hypothesized that injecting a small amount of fuel during negative valve overlap would allow unburned hydrocarbons in the internal residual to react, which could facilitate compression ignition during the following cycle.

In 2004, Yap et al. showed that while using internal EGR is promising for extending the load range and achieving the benefits of low NOx operation in gasoline engines, the same cannot be said for natural gas (NG) HCCI engines [81]. It was found that due to the energy requirements for NG autoignition, intake heating and high compression ratios are required to achieve autoignition in the NG HCCI engine. Internal EGR has the potential to reduce the intake heating requirement for NG combustion, but because of the high compression ratios necessary to achieve autoignition the amount of internal EGR available for mixture dilution was significantly reduced. In addition high combustion temperatures from NG HCCI combustion can lead to significantly higher emissions when compared to a gasoline HCCI engine [81].

Cairns and Blaxill combined the concepts of internal and external EGR to extend the load range of a multicylinder gasoline HCCI engine while avoiding knock [82]. It was also found that this combined EGR scheme could be used to facilitate a smooth transition between controlled autoignition (or HCCI) and SI modes, utilizing a hybrid combustion technique expanding the engine's operating range. Kawasaki et al. addressed some of these problems by experimenting with the opening of the intake valve (a small amount during the exhaust stroke). This "pilot opening" allows for exhaust gases to be pulled into the intake manifold, thus heating the intake mixture and increasing the total amount of internal EGR [83].

4.2. Changing Fuel Properties for HCCI Control

Changing fuel properties of the cylinder mixture is a method that can be used for HCCI control. The required time and conditions needed for autoignition vary between fuels, so combustion timing can be controlled and the operating range can be expanded by varying the fuel properties in an HCCI engine [71].

4.2.1. Modulating Two or More Fuels

Dual fuel usage is a method that can be used to actively vary the fuel octane number by mixing a fuel with a high octane number and a fuel with a low octane number to create a fuel mixture with an intermediate octane number. Furutani et al. were one of the first that combined two different fuels to control the autoignition timing [47]. A low octane fuel (n-heptane) was injected into a high octane homogeneous air/fuel mixture (propane or hydrogen) just before the intake valve. They found that more torque can be obtained by using fuels with more octane number differences. However, some amount of high-octane fuel does not participate in oxidation reactions because of its poor self-ignition tendency, so hydrocarbon emissions increase.

Stanglmaier et al. found that HCCI combustion timing could be controlled by mixing Fischer Tropsch (FT) Naptha with NG in an NG HCCI engine, allowing for optimization of efficiency and emissions at part loads [84]. Shibata et al. conducted a study on the effects of fuel properties on HCCI engine performance [85]. In this study fuels with different octane numbers were used in a four-cylinder engine. The resulting values of low temperature heat release (LTHR) and high temperature heat release (HTHR) varied with fuel composition. The low temperature chemical kinetics during LTHR as well as the negative temperature coefficient regime between LTHR and HTHR has been observed to have a large impact on HCCI combustion [84].

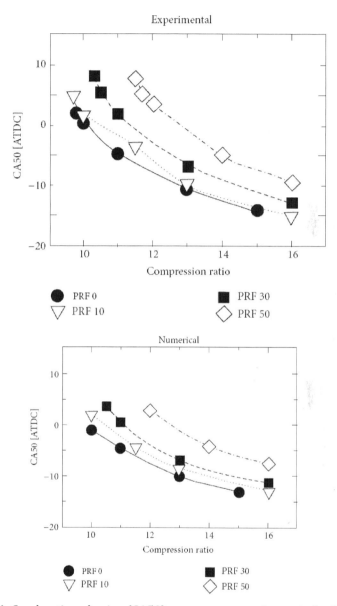

Figure 4. Combustion phasing (CA50) versus compression ratio for four PRF blends [49].

In an expanded study of HCCI control in 2007, Wilhelmsson et al. used dual fuels, NG and n-heptane, and a variable geometry turbocharger to develop an operational scheme in a NG engine by adding the lowest possible boost pressure to reduce pumping losses and minimize NOx emissions [86].

The effects of different primary reference fuel blends on HCCI operating range, start of combustion, burn duration, IMEP, indicated specific emissions, and indicated specific fuel consumption were investigated by Atkins and Koch who found that by changing the fuel octane number the HCCI operating range could be expanded [77]. Recently an experimental and numerical study has been performed by Dumitrescu et al. to determine the influence of isooctane addition on the combustion and emission characteristics of a HCCI engine fueled with n-heptane [49]. Results show that for the operating conditions studied (CR from 10 to 16, engine speed of 900 rpm, AFR 50, 30°C intake temperature, and no EGR), isooctane addition retarded the combustion phasing and reduced combustion efficiency. As shown in Figure 4, when compression ratio increased from 10 to 15, CA50 advanced 14 deg CA for PRF0, while CA50 advanced 17 deg CA for PRF50 when CR increased from 11.5 to 16. This suggests that a blend with more isooctane is more sensitive to compression ratio. Also the operating compression ratio range narrowed with increasing isooctane fraction in the fuel. The NOx emissions at advanced CA50 increased with increasing isooctane fraction, but the difference became negligible once CA50 approached TDC and beyond.

In 2004, Strandh et al. designed a PID controller and a model based linear quadratic Gaussian controller to establish cycle-by-cycle ignition timing control of an engine using blends of ethanol and n-heptane [87]. Dec and Berntsson separately found that a large amount of fuel stratification can lead to retarded ignition timing, which provides an additional actuator for control; however, too much stratification can ultimately lead to unstable combustion [88, 89].

4.2.2. Fuel Additives and Reforming

A potential technique for controlling the combustion timing of an HCCI engine is to change the fuel chemistry using two or more fuels with different autoignition attributes. Although a dual-fuel engine concept is technically achievable with current engine technologies, this is not ordinarily seen as a practical solution due to the indispensability of supplying and storing two fuels. Reformer gas (RG) is a combination of light gases dominated by hydrogen and carbon monoxide that can be produced from any hydrocarbon fuel using an onboard fuel processor. Reformer gas has the wide flammability limits and high resistance to autoignition [57].

Significant research exists on the addition of reformer gas to fuels of various compositions to control HCCI combustion, which is interesting because of the ability to produce reformer gas from other fuels, effectively eliminating the need for two separate fuel sources. As shown in Figure 5, the experimental study of Hosseini and Checkel demonstrates that increasing the reformer gas fraction retards the combustion timing to a more optimized value causing indicated power and fuel conversion efficiency to increase. Reformer gas reduces the first stage of heat release, extends the negative temperature coefficient delay period, and retards the main stage of combustion. In their study, two extreme cases of RG composition with H_2/CO ratios of 3/1 and 1/1 were investigated. The results demonstrate that both RG compositions retard the combustion phasing, but that the higher hydrogen fraction RG is more effective. Experimental work in this area has been completed by Hosseini and Checkel [90-93] and numerical works by Kongsereeparp and Checkel [94, 95].

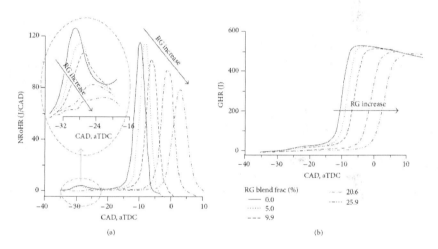

Figure 5. Effect of reformer gas on (a) net rate of heat release and (b) gross cumulative heat release [57].

4.3. Fast Thermal Management for HCCI Control

Fast Thermal Management (FTM) is a controlling technique that involves rapidly changing the temperature of intake charge to control the combustion phasing. Many studies have indicated that HCCI combustion timing is sensitive to intake air temperature [19, 42, 44, 96, 97]. Haraldsson et al. and Yang et al. suggested the use of two air streams and regaining heat from exhaust gases to heat one of the air streams [43, 98]. By mixing two air streams, one direct from atmosphere and the other heated by exhaust gases, it is possible to control the temperature of the final intake air stream (each stream with independent throttles for mixing). Both studies observed the ability of the FTM system to control the combustion phasing of HCCI combustion. The study by Yang indicates that while FTM is effective to control combustion phasing in HCCI engines, the "thermal inertia" of the system makes cycle by cycle temperature adjustment difficult, which in turn complicates the control of HCCI combustion during transients [98]. This lag in achieving the desired HCCI combustion phasing was also observed by Haraldsson research, although in that study FTM was presented as an acceptable alternative to use variable compression ratio in closed loop control of HCCI combustion [43].

4.3.1. Intake Temperature

The effects of intake charge temperature on HCCI combustion on-set have been widely reported by many researchers. In 1983, Najt and Foster showed that HCCI of lean mixtures could be achieved in a SI engine that has a low compression ratio with elevated intake charge temperatures (300–500°C) [19]. In general, the intake charge temperature has a strong influence on the HCCI combustion timing. Figure 6 demonstrates the combustion chamber pressure versus the crank angle for a 2-stroke engine at the speed of 6000 rpm [99]. As shown in this figure, increasing the overall gas temperature significantly advances the HCCI combustion timing. In temperature of 575 [°K], the ignition is so advanced and the combustion is not so efficient but by decreasing the temperature the ignition would be retarded. Also by decreasing the intake temperature the maximum pressure of cylinder decreases but at the intake temperature of 525 [°K] the ignition timing would be so retarded that causes some misfiring.

Figure 7 shows the NOx, CO, and HC emissions for various intake temperatures in the same engine. By decreasing the temperature and retarding the ignition timing, the NOx emission has decreased, but CO and HC emissions have increased. These adverse trends of CO and NOx emissions are one of the main difficulties for controlling the emissions since by reducing one of them, another one increases. Also as demonstrated in this figure, the trend of emissions at intake temperature of 525 [°K] has changed and NOx emission has suddenly increased because of some misfiring occurring in this point that was mentioned before.

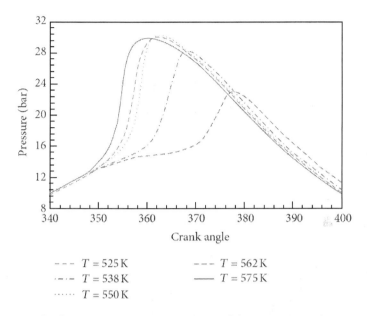

Figure 6. Cylinder pressure versus crank angel for various intake temperatures [99].

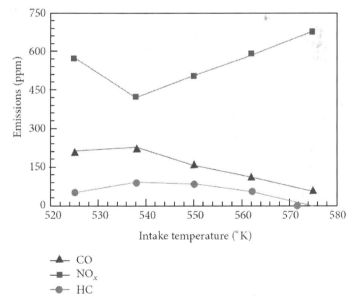

Figure 7. CO, NOx, and HC emissions for various intake temperatures [99].

The study performed by Iida and Igarashi also indicated that an increase in intake charge temperature (from 297°K to 355°K) increased the peak temperature after compression and advanced the HCCI combustion on-set [96]. Furthermore, the authors found that the effect of intake charge temperature on combustion on-set was greater for higher engine speed (1200 RPM) compared to the lower engine speed (600 RPM). Aceves and his coworkers carried out some investigations including analysis as well as experimental work [42]. On analysis, they developed two powerful tools: a single zone model and a multizone model. On experimental work, they did a thorough evaluation of operating conditions in a 4-cylinder Volkswagen TDI engine. The engine had been operated over a wide range of conditions by adjusting the intake temperature and the fuel flow rate. They found out that it may be possible to improve combustion efficiency by going to a lower fuel flow rate and a higher intake temperature. For the high load operating points, the trend was that lower intake temperature results in higher BMEP.

The effect of intake temperature on HCCI operation using negative valve overlap was investigated by Persson et al. [97]. They tested several points in the range between 15°C and 50°C to investigate the effects of intake charge temperature on spark assisted and unassisted HCCI combustion stabilities (COV_{IMEP} and) for a particular load and negative valve overlap condition. The study indicated that either increase in the residuals or intake charge temperature resulted in low coefficient of variation (COV) and stabilized the combustion. Recently, Mauyara and Agarwal experimentally investigated the effect of intake air temperature on cycle-to-cycle variations of HCCI combustion and performance parameters [44]. The cycle-to-cycle variations in combustion and performance parameters of HCCI combustion were investigated on a modified two cylinder direct injection diesel engine. The inlet air was supplied at 120, 140, and 160°C temperature. It was found that at lower intake air temperature it is possible to ignite the richer mixture (up to NOx) in HCCI combustion mode. As intake air temperature increases, engine running on richer mixture tends to knock with very high rate of pressure rise. But at higher intake air temperature it is possible to ignite the leaner mixture (up to NOx) in HCCI combustion mode.

4.3.2. Compression Ratio

Compression ratio as an effective means to achieve HCCI combustion control has been carefully investigated by Christensen et al. for several years [33, 69]. His studies demonstrated that regardless of fuel type used increasing the compression ratio (9.6 : 1–22.5 : 1) had a strong influence on ignition timing and assists in decreasing the necessary intake charge temperature. Hiraya et al. also reported the effect of compression ratio (12 : 1–18.6 : 1) on combustion on-set [100]. Their study on a gasoline HCCI engine showed that higher compression ratios allowed for lower intake charge temperature and higher intake density for higher output. Furthermore, higher compression ratio contributed to higher thermal efficiency. The study done by Iida also has confirmed that change in compression ratio has a strong influence on HCCI combustion on-set [96]. Their results also showed that compression ratio has a greater effect on HCCI combustion on-set compared to changes in either intake charge temperature or coolant temperature.

The study done by Olsson et al. investigated the influence of compression ratio on a natural gas fuelled HCCI engine [34]. The experimental engine had a secondary piston that was installed in the cylinder head whose position can be varied to attain variable compression ratio (VCR). In their tests, the compression ratio was modified (21:1, 20:1, 17:1, and 15:1) according to the operating condition to attain autoignition of the charge close to TDC. This VCR engine showed the potential to achieve satisfactory operation in HCCI mode over a wide range of operating conditions by using the optimal compression ratio for a particular operating condition. The study also showed that the maximum pressure rise rate increased with higher compression ratio for early combustion timing and a reverse effect was seen with delayed combustion on-set.

Haraldsson et al. investigated HCCI combustion phasing with closed-loop combustion control using variable compression ratio in a multicylinder engine [36]. In his study, closed-loop combustion control using accurate and fast variable compression ratio was run with acceptable performance. Time constant of three engine cycles was achieved for the compression ratio control. The closed-loop combustion control system of cascade coupled compression ratio and CA50 controllers had a time constant of 14 engine cycles or 0.84 s at 2000 rpm with a dCA50/dt of 6.0 CAD/s.

4.4. Direct Injection for HCCI Control

Fuel injection into the cylinder at different stages of the engine cycle allows HCCI combustion timing to be advanced by improving mixture ignitability or retarded by increasing fuel stratification, creating the possibility of expanding the low and high load operating limits. Direct injection can be a good way to control HCCI combustion, but it depends heavily on the type of fuel and the timing of the direct injection [71].

A numerical study by Gong et al. showed that power density of an HCCI engine could be improved by the injection of a small amount of diesel fuel during the compression stroke of the engine. This pilot fuel injection also decreased the sensitivity of the HCCI combustion to intake conditions [101].

In 2003, Wagner et al. demonstrated that it would not be possible to use n-heptane as a port injection fuel for HCCI and instead a carefully timed n-heptane direct cylinder injection is used to avoid wall impingement and utilize the benefits of HCCI combustion [102]. In that year, Urushihara et al. found that a small injection of fuel during the NVO interval and a second injection during the intake stroke result in internal fuel reformation, which improves the ignitability of the cylinder mixture [103].

Dec and Sjöberg found that direct injection of fuel early in the intake stroke produced near identical results to a premixed charge. However, injection close to TDC improved the combustion efficiency of very low fuel load mixtures [104]. Numerical models by Strålin et al. showed that fuel stratification caused by injection of fuel around TDC results in pockets of rich fuel and air mixture, which promotes ignitability. Overall fuel stratification extended the combustion duration helping to avoid knock, thus extending the operating range of the engine [105]. Helmantel and Denbratt used multiple injection scheme of n-heptane to allow for sufficient mixing to operate a conventional diesel common passenger rail car engine with HCCI combustion [106].

In agreement with the recent study of Lu et al., for stratified charge compression ignition (SCCI) combustion with Port Fuel Injection of the two-stage reaction fuel combined with in-cylinder direct injection, the heat release rate demonstrates a three-stage heat release, as shown in Figure 8 [53]. The combustion phasing and the peak value of first-stage combustion play a vital role in the ignition timing and the peak point of the second-stage combustion, while the crucial factors of the first-stage reaction are the chemical properties of the premixed fuel. The second-stage ignition timing and peak point have an important influence on the combustion phasing of the third-stage combustion, the thermal efficiency, the maximum gas temperature, and the knock intensity or the pressure rise rate. The dominant factors of the second-stage reaction are the premixed ratio and the physical properties of the premixed fuel. The third-stage combustion controls the engine thermal efficiency, the overall combustion efficiency, and NO_x and other emissions. Its decisive factor is the in-cylinder injection timing. If the ignition timing and peak value of each stage reaction can be flexibly dominated using mixture concentration stratification, composition stratification, and temperature stratification, then the expanded engine load, optimized thermal efficiency, and lowest NO_x emissions may be achieved [53].

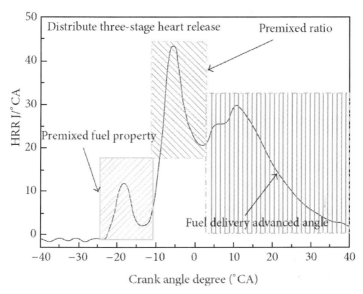

Figure 8. Three-stage heat release and its influential factors of SCCI combustion [53].

Recently Yang et al. did an experimental study of fuel stratification for HCCI high load extension [51]. The investigation was performed in a single-cylinder four-stroke engine equipped with a dual fuel injection system, a port injector for preparing a homogeneous charge with gasoline and a direct in-cylinder injector for creating the desired fuel stratification with gasoline or methanol. Both the effect of gasoline fuel stratification and gasoline/methanol stratification were parametrically investigated. Test results indicated that weak gasoline stratification leads to an advanced combustion phase and an increase in NO_x emission, while increasing the stratification with a higher quantity of gasoline direct injection results in a significant deterioration in both the combustion efficiency and the CO emission. Engine tests using methanol for the stratification retarded the ignition timing and prolonged the combustion duration, resulting in a substantial reduction in the maximum rate of pressure rise and the maximum cylinder pressure a prerequisite for HCCI high load extension. About the stratified methanol-to-gasoline compared to gasoline HCCI, a 50% increase in the maximum IMEP attained was achieved with an acceptable maximum pressure rise rate of 0.5 MPa/°CA while maintaining a high thermal efficiency [51].

5. CONCLUSION

CAI/HCCI engines still have not met the level of development and cost that would make a market introduction possible at the moment. The technical challenges facing both gasoline and diesel HCCI combustion are their limited operational range and less optimized combustion phasing, owing to the lack of direct control over the start of ignition and the rate of heat release. HCCI combustion represents a step change in combustion technology and its future research and application should be considered as part of an effort to achieve low-temperature combustion in a wide range of operating conditions in an IC engine. Combustion process in future IC engines converges towards premixed compression ignition combustion, while turbocharging and direct injection become a norm on such engines: it therefore may not remain futuristic but become a realistic possibility that, with more flexible engine hardware and their real-time control, a fully flexible engine could be developed to convert the chemical energy from any type of fuel into mechanical work through premixed auto-ignited low-temperature combustion [1].

ACKNOWLEDGMENT

The author acknowledges the support of Universiti Putra Malaysia under Research University Grants (RUGS), Project no. 05-05-10-1076RU and Ministry of Higher Education under Exploratory Research Grants Scheme (ERGS), Project Code: ERGS/1/2012/TK01/UPM/02/5 for this research.

REFERENCES

1. H. Zhao, HCCI and CAI Engines for the Automotive Industry, Woodhead Publishing Limited, Cambridge, UK, 2007.
2. F. Zhao, T. N. Asmus, D. N. Assanis, J. E. Dec, J. A. Eng, and P. M. Najt, "Homogeneous charge compression ignition (HCCI) engines," SAE Publication, 2003.
3. R. H. Stanglmaier and C. E. Roberts, "Homogeneous charge compression ignition (HCCI): benefits, compromises, and future engine applications," SAE Paper 1999-01-3682, 1999.

4. M. Shahbakhti, Modeling and experimental study of an HCCI engine for combustion timing control [Ph.D. thesis], University of Alberta, 2009.

5. A. Hakansson, CA50 estimation on HCCI engine using engine speed variations [MSc thesis], Lund University, 2007.

6. O. Erlandsson, "Early swedish hot-bulb engines—efficiency and performance compared to contemporary gasoline and diesel engines," SAE Paper 2002-01-0115, 2002.

7. S. H. Jo, P. D. Jo, T. Gomi, and S. Ohnishi, "Development of a low-emission and high-performance 2-stroke gasoline engine (NiCE)," SAE Paper 730463, 1973.

8. S. Onishi, S. H. Jo, K. Shoda, P. D. Jo, and S. Kato, "Active thermo-atmosphere combustion (ATAC)—a new combustion process for internal combustion engines," SAE Paper 790501, 1979.

9. M. Noguchi, Y. Tanaka, T. Tanaka, and Y. Takeuchi, "A study on gasoline engine combustion by observation of intermediate reactive products during combustion," SAE Paper 790840, 1979.

10. K. Tsuchiya, S. Hirano, M. Okamura, and T. Gotoh, "Emission control of two-stroke motorcycle engines by the butterfly exhaust valve," SAE Paper 800973, 1980.

11. P. Duret and J. F. Moreau, "Reduction of pollutant emissions of the IAPAC two-stroke engine with compressed air assisted fuel injection," SAE Paper 900801, 1990.

12. P. Duret and S. Venturi, "Automotive calibration of the IAPAC fluid dynamically controlled two-stroke combustion process," SAE Paper 960363, 1996.

13. Y. Ishibashi and Y. Tsushima, "A trial for stabilizing combustion in two-stroke engines at part throttle operation," in Proceedings of the IFP International Seminar, Editions Technip, Rueil-Malmaison, France, 1993.

14. M. Asai, T. Kurosaki, and K. Okada, "Analysis on fuel economy improvement and exhaust emission reduction in a two-stroke engine by using an exhaust valve," SAE Paper 951764, 1995.

15. Y. Ishibashi and M. Asai, "Improving the exhaust emissions of two-stroke engines by applying the activated radical combustion," SAE Paper 960742, 1996.

16. P. R. Hooper, T. Al-Shemmeri, and M. J. Goodwin, "Advanced modern low-emission two-stroke cycle engines," Journal of Automobile Engineering, vol. 225, no. 11, pp. 1531–1543, 2011.

17. D. W. Blundell, J. Turner, R. Pearson, R. Patel, and J. Young, "The omnivore wide-range auto-ignition engine: results to date using 98RON unleaded gasoline and E85 fuels," SAE Paper 2010-01-0846, 2010.

18. J. W. G. Turner, D. W. Blundell, and R. J. Pearson, "Project omnivore: a variable compression ratio ATAC 2-stroke engine for ultra-wide-range hcci operation on a variety of fuels," SAE Paper 2010-01-1249, 2010.

19. P. M. Najt and D. E. Foster, "Compression-ignited homogeneous charge combustion," SAE Paper 830264, 1983.

20. R. H. Thring, "Homogeneous-charge compression-ignition (HCCI) engines," SAE Paper 892068, 1989.

21. J. O. Olsson, P. Tunestål, and B. Johansson, "Closed-loop control of an hcci engine," SAE Paper 2001-01-1031, 2001.

22. M. Stockinger, H. Schäpertöns, and P. Kuhlmann, Versuche an Einem Gemischansaugenden mit Selbszündung, MTZ, 1992.

23. J. O. Olsson, O. Erlandsson, and B. Johansson, "Experiments and simulation of a six-cylinder homogeneous charge compression ignition (HCCI) engine," SAE Paper 2000-01-2867, 2000.

24. D. S. Kim and C. S. Lee, "Improved emission characteristics of HCCI engine by various premixed fuels and cooled EGR," Fuel, vol. 85, no. 5-6, pp. 695–704, 2006.

25. A. Dubreuil, F. Foucher, C. Mounaim-Rousselle, G. Dayma, and P. Dagaut, "HCCI combustion: effect of NO in EGR," Proceedings of the Combustion Institute, vol. 31, pp. 2879–2886, 2007.

26. L. Manofsky, J. Vavra, D. Assanis, and A. Babajimopoulos, "Bridging the gap between HCCI and SI: spark-assisted compression ignition," SAE Paper, no. 2011-01-1179, 2011.

27. Q. L. Nguyen, The effects of operating parameters on combustion and emissions of si engine—a pre study of hcci combustion [MSc thesis], Southern Taiwan University, 2007.

28. D. Law, D. Kemp, J. Allen, G. Kirkpatrick, and T. Coplan, "Controlled combustion in an IC-engine with a fully variable valve train," SAE Paper 2001-01-0251, 2001.

29. M. Jennische, Closed-loop control of start of combustion in a homogeneous charge compression ignition engine [MSc thesis], Lund Institute of Technology, 2003.

30. P. A. Caton, A. J. Simon, J. C. Gerdes, and C. F. Edwards, "Residual-effected homogeneous charge compression ignition at a low compression ratio using exhaust reinduction," International Journal of Engine Research, vol. 4, pp. 163–177, 2003.

31. S. Yamaoka, H. Kakuya, S. Nakagawa, T. Okada, A. Shimada, and Y. Kihara, "HCCI operation control in a multi-cylinder gasoline engine," SAE Paper 2005-01-0120, 2005.

32. F. Agrell, H. E. Ångström, B. Eriksson, J. Wikander, and J. Linderyd, "Integrated simulation and engine test of closed loop hcci control by aid of variable valve timings," SAE Paper 2003-01-0748, 2003.

33. M. Christensen, A. Hultqvist, and B. Johansson, "Demonstrating the multi fuel capability of a homogeneous charge compression ignition engine with variable compression ratio," SAE Paper 1999-01-3679, 1999.

34. J. O. Olsson, P. Tunestål, B. J. Johansson, S. Fiveland, R. Agama, and M. Willi, "Compression ratio influence on maximum load of a natural gas fueled HCCI engine," SAE Paper 2002-01-0111, 2002.

35. G. Haraldsson and B. Johansson, "Supercharging HCCI to extend the operating range in a multi-cylinder VCR HCCI engine," SAE Paper 2003-01-3214, 2003.

36. G. Haraldsson, P. Tunestål, and B. Johansson, "HCCI combustion phasing with closed-loop combustion control using variable compression ratio in a multi cylinder engine," SAE Paper 2003-01-1830, 2003.

37. R. Chen, N. Milovanovic, J. Turner, and D. Blundell, "The thermal effect of internal exhaust gas recirculation on controlled auto ignition," SAE Paper 2003-01-0751, 2003.

38. M. Fathi, R. K. Saray, and M. D. Checkel, "The influence of Exhaust Gas Recirculation (EGR) on combustion and emissions of n-heptane/natural gas fueled Homogeneous Charge Compression Ignition (HCCI) engines," Applied Energy, vol. 88, no. 12, pp. 4719–4724, 2011.

39. Y. Takeda, N. Keiichi, and N. Keiichi, "Emission characteristics of premixed lean diesel combustion with extremely early staged fuel injection," SAE Paper 961163, 1996.

References

40. K. Nakagome, N. Shimazaki, K. Niimura, and S. Kobayashi, "Combustion and emission characteristics of premixed lean diesel combustion engine," SAE Paper 970898, 1997.

41. D. I. Handford and M. D. Checkel, "Extending the load range of a natural gas HCCI engine using direct injected pilot charge and external EGR," SAE Paper 2009-01-1884, 2009.

42. S. M. Aceves, D. L. Flowers, J. Martinez-Frias, and J. R. Smith, "HCCI combustion: analysis and experiments," SAE Paper 2001-01-2077, 2001.

43. G. Haraldsson, P. Tunestål, B. Johansson, and J. Hyvönen, "HCCI closed-loop combustion control using fast thermal management," SAE Paper 2004-01-0943, 2004.

44. R. K. Maurya and A. K. Agarwal, "Experimental investigation on the effect of intake air temperature and air-fuel ratio on cycle-to-cycle variations of HCCI combustion and performance parameters," Applied Energy, vol. 88, no. 4, pp. 1153–1163, 2011.

45. M. Christensen and B. Johansson, "Homogeneous charge compression ignition with water injection," SAE Paper 1999-01-0182, 1999.

46. N. Milovanovic, D. Blundell, R. Pearson, J. Turner, and R. Chen, "Enlarging the operational range of a gasoline HCCI engine by controlling the coolant temperature," SAE Paper 2005-01-0157, 2005.

47. M. Furutani, Y. Ohta, M. Kovo, and M. Hasegawa, "An ultra-lean premixed compression ignition engine concept and its characteristics," in Proceedings of the 4th International Symposium COMODIA, 1998.

48. H. Akagawa, T. Miyamoto, A. Harada, S. Sasaki, N. Shimazaki, and T. Hashizume, "Approaches to solve problems of the premixed lean diesel combustion," SAE Paper 1999-01-0183, 1999.

49. C. E. Dumitrescu, H. Guo, V. Hosseini et al., "The effect of iso-octane addition on combustion and emission characteristics of a HCCI engine fueled with n-heptane," Journal of Engineering for Gas Turbines and Power, vol. 133, no. 11, Article ID 112801, 2011.

50. Z. Zheng and M. Yao, "Charge stratification to control HCCI: experiments and CFD modeling with n-heptane as fuel," Fuel, vol. 88, no. 2, pp. 354–365, 2009.

51. D.-B. Yang, Z. Wang, J.-X. Wang, and S.-J. Shuai, "Experimental study of fuel stratification for HCCI high load extension," Applied Energy, vol. 88, no. 9, pp. 2949–2954, 2011.

52. Y. Yang, J. E. Dec, N. Dronniou, and M. Sjöberg, "Tailoring HCCI heat-release rates with partial fuel stratification: comparison of two-stage and single-stage-ignition fuels," Proceedings of the Combustion Institute, vol. 33, pp. 3047–3055, 2011.

53. X. Lu, Y. Shen, Y. Zhang et al., "Controlled three-stage heat release of stratified charge compression ignition (SCCI) combustion with a two-stage primary reference fuel supply," Fuel, vol. 90, no. 5, pp. 2026–2038, 2011.

54. H. Liu, Z. Zheng, M. Yao et al., "Influence of temperature and mixture stratification on HCCI combustion using chemiluminescence images and CFD analysis," Applied Thermal Engineering, vol. 33-34, no. 1, pp. 135–143, 2012.

55. D. Yap, A. Megaritis, S. Peucheret, M. Wyszynski, and H. Xu, "Effect of hydrogen addition on natural gas HCCI combustion," SAE Paper 2004-01-1972, 2004.

56. H. Zu, T. Wilson, S. Richardson, M. Wyszynski, T. Megaritis, D. Yap, et al., "Extension of the boundary of HCCI combustion using fuel reforming technology," in JSAE Annual Congress, pp. 23–26, 2004.

57. V. Hosseini, W. Stuart Neill, and M. David Checkel, "Controlling n-heptane HCCI combustion with partial reforming: experimental results and modeling analysis," Journal of Engineering for Gas Turbines and Power, vol. 131, no. 5, Article ID 052801, 2009.

58. F. J. Martinez, S. M. Aceves, D. Flowers, J. R. Smith, and R. Dibble, "Equivalence ratio-EGR control of HCCI engine operation and the potential for transition to spark-ignited operation," SAE Paper 2001-01-3613, 2001.

59. M. Christensen, B. Johansson, P. Amnéus, and F. Mauss, "Supercharged homogeneous charge compression ignition," SAE Paper 980787, 1998.

60. Y. Iwabuchi, K. Kawai, T. Shoji, and Y. Takeda, "Trial of New concept diesel combustion system—premixed compression-ignited combustion," SAE Paper 1999-01-0185, 1999.

61. R. Sun, R. Thomas, and C. L. Gray, "An HCCI Engine: power plant for a hybrid vehicle," SAE Paper 2004-01-0933, 2004.

62. J. O. Olsson, P. Tunestål, G. Haraldsson, and B. Johansson, "A turbo charged dual fuel HCCI engine.,"SAE Paper 2001-01-1896, 2001.

63. J. O. Olsson, P. Tunestål, and B. Johansson, "Boosting for high load HCCI," SAE Paper 2004-01-0940, 2004.

64. L. Koopmans, H. Ström, S. Lundgren, O. Backlund, and I. Denbratt, "Demonstrating a SI-HCCI-SI mode change on a Volvo 5-cylinder electronic valve control engine," SAE Paper 2003-01-0753, 2003.

65. H. Santoso, J. Matthews, and W. K. Cheng, "Managing SI/HCCI dual-mode engine operation," SAE Paper 2005-01-0162, 2005.

66. M. Canova, F. Chiara, J. Cowgill, M. S. Midlam, Y. Guezennec, and G. Rizzoni, "Experimental characterization of mixed-mode HCCI/DI combustion on a common rail diesel engine," SAE Paper 2007-24-0085, 2007.

67. J. L. Burton, D. R. Williams, W. J. Glewen, M. J. Andrie, R. B. Krieger, and D. E. Foster, "Investigation of transient emissions and mixed mode combustion for a light duty diesel engine," SAE Paper 2009-01-1347, 2009.

68. T. Aoyama, Y. Hattori, J. Mizuta, and Y. Sato, "An experimental study on premixed-charge compression ignition gasoline engine," SAE Paper 960081, 1996.

69. M. Christensen, B. Johansson, and P. Einewall, "Homogeneous charge compression ignition (HCCI) using isooctane, ethanol and natural gas—a comparison with spark ignition operation," SAE Paper 972874, 1997.

70. A. Harada, N. Shimazaki, S. Sasaki, T. Miyamoto, H. Akagawa, and K. Tsujimura, "The effects of mixture formation on premixed lean diesel combustion engine," SAE Paper 980533, 1998.

71. M. N. Schleppe, SI-HCCI mode switching optimization using a physics based model [M.S. thesis], University of Alberta, 2011.

72. M. Christensen and B. Johansson, "Influence of mixture quality on homogeneous charge compression ignition," SAE Paper 982454, 1998.

73. P. Zoldak, Design of a research engine for homogeneous charge compression ignition (HCCI) combustion [M.S. thesis], University of Windsor, 2005.

74. M. Christensen and B. Johansson, "Supercharged homogeneous charge compression ignition (HCCI) with exhaust gas recirculation and pilot fuel," SAE Paper 2000-01-1835, 2000.

75. S. S. Morimoto, Y. Kawabata, T. Sakurai, and T. Amano, "Operating characteristics of a natural gas-fired homogeneous charge

compression ignition engine (performance improvement using EGR)," SAE Paper2001-01-1034, 2001.

76. K. Narayanaswamy and C. J. Rutland, "Cycle simulation diesel HCCI modeling studies and control," SAE Paper 2004-01-2997, 2004.

77. M. J. Atkins and C. R. Koch, "The effect of fuel octane and dilutent on homogeneous charge compression ignition combustion," Journal of Automobile Engineering, vol. 219, no. 5, pp. 665–675, 2005.

78. H. Zhao, Z. Peng, J. Williams, and N. Ladommatos, "Understanding the effects of recycled burnt gases on the controlled autoignition (CAI) combustion in four-stroke gasoline engines," SAE Paper 2001-01-3607, 2001.

79. N. Milovanovic, R. Chen, and J. Turner, "Influence of the variable valve timing strategy on the control of a homogeneous charge compression (HCCI) engine," SAE Paper 2004-01-1899, 2004.

80. Y. Urata, M. Awasaka, J. Takanashi, T. Kakinuma, T. Hakozaki, and A. Umemoto, "A study of gasoline-fuelled HCCI engine equipped with an electromagnetic valve train," SAE Paper 2004-01-1898, 2004.

81. D. Yap, A. Megaritis, M. L. Wyszynski, and H. Xu, "residual gas trapping for natural gas HCCI," SAE Paper 2004-01-1973, 2004.

82. A. Cairns and H. Blaxill, "The effects of combined internal and external exhaust gas recirculation on gasoline controlled auto-ignition," SAE Paper 2005-01-0133, 2005.

83. K. Kawasaki, A. Takegoshi, K. Yamane, H. Ohtsubo, T. Nakazono, and K. Yamauchi, "Combustion improvement and control for a natural gas HCCI engine by the internal EGR by means of intake-valve pilot-opening," SAE Paper 2006-01-0208, 2006.

84. R. H. Stanglmaier, T. W. Ryan, and J. S. Souder, "HCCI operation of a dual-fuel natural gas engine for improved fuel efficiency and ultra-low NOx emissions at low to moderate engine loads," SAE Paper 2001-01-1897, 2001.

85. G. Shibata, K. Oyama, T. Urushihara, and T. Nakano, "The effect of fuel properties on low and high temperature heat release and resulting performance of an HCCI engine," SAE Paper 2004-01-0553, 2004.

86. C. Wilhelmsson, P. Tunestảa, and B. Johansson, "Operation strategy of a dual fuel HCCI engine with VGT," SAE Paper 2007-01-1855, 2007.

87. P. Strandh, J. Bengtsson, R. Johansson, P. Tunestål, and B. Johansson, "Cycle-to-cycle control of a dual-fuel HCCI engine," SAE Paper 2004-01-0941, 2004.
88. J. E. Dec and M. Sjöberg, "Isolating the effects of fuel chemistry on combustion phasing in an HCCI engine and the potential of fuel stratification for ignition control," SAE Paper 2004-01-0557, 2004.
89. A. W. Berntsson and I. Denbratt, "HCCI combustion using charge stratification for combustion control," SAE Paper 2007-01-0210, 2007.
90. V. Hosseini and M. D. Checkel, "Using reformer gas to enhance HCCI combustion of CNG in a CFR engine," SAE Paper 2006-01-3247, 2006.
91. V. Hosseini and M. D. Checkel, "Effect of reformer gas on HCCI combustion—part I: high octane fuels," SAE Paper 2007-01-0208, 2007.
92. V. Hosseini and M. D. Checkel, "Effect of reformer gas on HCCI combustion—part II: low octane fuels," SAE Paper 2007-01-0206, 2007.
93. V. Hosseini and M. D. Checkel, "reformer gas composition effect on HCCI combustion of n-heptane, iso-octane, and natural gas," SAE Paper 2008-01-0049, 2008.
94. P. Kongsereeparp and M. D. Checkel, "Environmental, thermodynamic and chemical factor effects on heptane-and CNG-fuelled HCCI combustion with various mixture compositions," SAE Paper 2008-01-0038, 2008.
95. P. Kongsereeparp and M. D. Checkel, "Study of reformer gas effects on n-heptane HCCI combustion using a chemical kinetic mechanism optimized by genetic algorithm," SAE Paper 2008-01-0039, 2008.
96. N. Iida and T. Igarashi, "Auto-ignition and combustion of n-butane and DME/air mixtures in a homogeneous charge compression ignition engine," SAE Paper 2000-01-1832, 2000.
97. H. Persson, M. Agrell, J. O. Olsson, B. Johansson, and H. Ström, "The effect of intake temperature on HCCI operation using negative valve overlap," SAE Paper 2004-01-0944, 2004.
98. J. Yang, T. Culp, and T. Kenney, "Development of a gasoline engine system using HCCI technology—the concept and the test results," SAE Paper 2002-01-2832, 2002.

99. M. Izadi Najafabadi, A. A. Nuraini, A. Nor Mariah, and L. Abdul Mutalib, "Effects of intake temperature and equivalence ratio on HCCI ignition timing and emissions of a 2-stroke engine," Applied Mechanics and Materials Journal, vol. 315, pp. 498–502, 2013.

100. K. Hiraya, K. Hasegawa, T. Urushihara, A. Iiyama, and T. Itoh, "A Study on gasoline fueled compression ignition engine—a trial of operation region expansion," SAE Paper 2002-01-0416, 2002.

101. W. Gong, S. R. Bell, G. J. Micklow, S. B. Fiveland, and M. L. Willi, "Using pilot diesel injection in a natural gas fueled HCCI engine," SAE Paper 2002-01-2866, 2002.

102. U. Wagner, R. Anca, A. Velji, and U. Spicher, "An experimental study of homogeneous charge compression ignition (HCCI) with various compression ratios, intake air temperatures and fuels with port and direct fuel injection," SAE Paper 2003-01-2293, 2003.

103. T. Urushihara, K. Hiraya, A. Kakuhou, and T. Itoh, "Expansion of HCCI operating region by the combination of direct fuel injection, negative valve overlap and internal fuel reformation," SAE Paper2003-01-0749, 2003.

104. J. E. Dec and M. Sjöberg, "A parametric study of HCCI combustion - the sources of emissions at low loads and the effects of GDI fuel injection," SAE Paper 2003-01-0752, 2003.

105. P. Strålin, F. Wåhlin, and H. E. Ångström, "Effects of injection timing on the conditions at top dead center for direct injected HCCI," SAE Paper 2003-01-3219, 2003.

106. A. Helmantel and I. Denbratt, "HCCI operation of a passenger car common rail DI diesel engine with early injection of conventional diesel fuel," SAE Paper 2004-01-0935, 2004.

CHAPTER 5

NO_x Storage and Reduction for Diesel Engine Exhaust Aftertreatment

Beñat Pereda-Ayo and Juan R. González-Velasco

1 Department of Chemical Engineering, Faculty of Science and Technology, University of the Basque Country UPV/EHU, Bilbao, Spain

1. INTRODUCTION

Diesel and lean-burn engines provide better fuel economy and produce lower CO_2 emissions compared to conventional Otto gasoline engines. However, the NO_x gas components in the lean (oxidizing) exhausts from diesel and lean-burn engines cannot be efficiently removed with the classical three-way catalyst (TWC) under operating conditions with excess of oxygen in the exhaust gas. Among the available technologies under research, the NO_x storage-reduction (NSR) catalyst seems to be the most promising method to solve the problem. Basically, NSR catalysts consist of a cordierite monolith washcoated with porous alumina on which an alkali or alkali-earth oxide (e.g. BaO) and a noble metal (Pt) are deposited. These catalysts operate under cyclic conditions. During the lean period, when oxygen is in excess, the platinum oxidizes NO to a mixture of NO and NO_2 (NO_x), which is adsorbed (stored) on Ba as various NO_x species (nitrate, nitrite). During the subsequent short rich period, when some reductant (e.g. H_2) is injected, NO_x ad-species are released and reduced to nitrogen on Pt. Ammonia and N_2O byproduct formation upon NO_x reduction can also be observed over Pt-BaO/Al_2O_3 NSR catalysts.

In this chapter a systematic methodology for preparing Pt-Ba/Al_2O_3 NSR monolith catalysts is presented, the NO_x storage and reduction mechanisms on the catalyst are analysed, and the optimal control of different operational variables to achieve the NSR process with maximum production and selectivity to nitrogen is modeled [1].

2. HISTORICAL BACKGROUND

Main pollutants generated in the engine exhaust gases are nitrogen oxides (NO_x), carbon oxides (CO_x), hydrocarbons (HC) and particulate matter (PM). The last term is referred to small particles leaving the engine, mainly constituted by carbonaceous material. These fine particles can enter into the human lungs, being responsible for some breathing and cardiovascular diseases [2].

The hydrocarbons are organic volatile compounds able to form ozone smog at the ground level when interacting with nitrogen oxides under the sun light. Ozone irritates the eyes, hurts the lungs, causes asthma attack and aggravates other respiratory problems. In addition, ozone is one of the primary components of photochemical smog (or just smog for short). Furthermore, hydrocarbons can also cause cancer [3].

Nitrogen oxides, same as hydrocarbons, are precursors for ozone formation. The NO_2 contributes importantly to the formation of acid rain [4]. The carbon monoxide (CO) reduces the oxygen flow in the blood and results particularly dangerous for people with heart diseases [5]. The carbon dioxide (CO_2) is a greenhouse gas able to make an atmosphere layer trapping the heat and contributing to the global warm of the earth [6].

2.1. Legislation

The negative impacts of those emissions on the human health and the environment and climate have forced legislation to control and limit such emissions. In the U.S.A., NO_x emissions from mobile sources contribute almost 50 % of those produced in total, so that more and stricter regulations have been introduced for reducing NO_x emissions from the automobiles [7].

Table 1 shows the most important regulations as introduced by the European Union from the first directive Euro 1 (1992), then Euro 2 (1996), Euro 3 (2000), Euro 4 (2005), Euro 5 (2009), and the most recent Euro 6 (2014) [8]. Emission limits for CO, HC, NO_x, and PM were proposed for petrol and diesel engines. Former regulations limited HC+NO_x jointly, which later were split up into individual HC and NO_x limits.

Table 1. EU emission standards for passenger cars, g km^{-1}.

Step	Date	CO	HC	HC+NO$_x$	NO$_x$	PM
Diesel engines						
Euro 1	07.1992	2.72	–	0.97	–	0.14
Euro 2	01.1996	1.00	–	0.70	–	0.08
Euro 3	01.2000	0.64	–	0.56	0.50	0.05
Euro 4	01.2005	0.50	–	0.30	0.25	0.025
Euro 5	09.2009	0.50	–	0.23	0.18	0.005
Euro 6	09.2014	0.50	–	0.17	0.08	0.005
Petrol engines						
Euro 1	07.1992	2.72	–	0.97	–	–
Euro 2	01.1996	2.20	–	0.50	–	–
Euro 3	01.2000	2.30	0.20	–	0.15	–
Euro 4	01.2005	1.00	0.10	–	0.08	–
Euro 5	09.2009	1.00	0.10	–	0.06	0.005
Euro 6	09.2014	1.00	0.10	–	0.06	0.005

On the other hand, Euro standards have been completed with stricter regulations for sulphur content in fuels. In fact, the content of S in diesel could not surpass 350 ppm from the year 2000, and only 50 ppm from 2005 (for petrol, 150 ppm in 2000 and 50 ppm in 2005). From 2009, S-free fuels (S ≤ 10 ppm) have been implemented.

In the most recent Euro standards the durability of the catalyst is also specified, e.g. Euro 3 required the emission standards for 80,000 km or 5 years (whatever first occurs). Following regulations required 100,000 km or 5 years. From 2000, with the entrance of Euro 3, vehicles should be equipped with on board diagnostics (OBD), announcing to the driver the system damage or wrong operation, then causing higher emissions which should be avoided.

2.2. Automobile Exhaust Aftertreatment

More exigent legislation on automobile exhaust emissions has led to the development of aftertreatment systems. Today, three way catalysts (TWC) oxidize CO and HC to CO_2 and H_2O, and simultaneously reduce NO_x to N_2 in a very efficient way for conventional Otto gasoline engines [9-12]. The conventional gasoline engines operate at stoichiometric air/fuel ratio, A/F=14.63 (w/w) [13,14], which produces an exhaust gas with the exact balance of CO, H_2 and HC (reducing species) needed to reduce NO_x and O_2 (oxidizing species). However, diesel engines operate with higher A/F ratios, from 20:1 to 65:1 [15], then producing an exhaust gas with oxygen in excess (Table 2 [16]).

Table 2. Exhaust gas composition, depending on the type of engine.

		Conventional gasoline engine	Diesel engine	Lean engine
O_2	%vol.	0.2 – 2	5 – 15	4 – 18
CO_2	%vol.	10 – 13.5	2 – 12	2 – 12
H_2O	%vol.	10 – 12	2 – 10	2 – 12
N_2	%vol.	70 – 75	70 – 75	70 – 75
CO	%vol.	0.1 – 6	0.01 – 0.1	0.04 – 0.08
HC	%vol. C_1	0.5 – 6	0.005 – 0.05	0.002 – 0.015
NO_x	%vol.	0.04 – 0.4	0.003 – 0.06	0.01 – 0.05
SO_x	Related to the content of S in the fuel			

Fig. 1 shows the conversion curves for each pollutant as a function of the air/fuel ratio, for a TWC. Around the stoichiometric point (A/F=14.63), all the three pollutants (HC, CO and NO) are highly converted (>95 %), i.e. they are almost totally removed. However, when the environment is abundant in oxygen as in diesel engines (A/F>20), although this environment enhances the oxidation of HC and CO, the reduction of NO becomes practically inefficient, then this pollutant cannot be appropriately removed with TWC technology [15,17,18].

On the other hand, technical solutions existing for the optimal compromise in removal of NO_x/PM [19], by exhaust gas recirculation (EGR), are not able to achieve the requirements of Euro 6. In fact, in these systems, reduction of PM means eventually an increment of NO_x and viceversa [20]. Consequently, current technologies combining diesel particulate filters (DPF) and $DeNO_x$ catalysts [21-23] are being reconsidered.

At present, the removal of NO_x in the diesel engine exhaust gases, mainly in heavy-duty lorries, is controlled by selective catalytic reduction (SCR) with ammonia generated by hydrolysis of urea which must be stored in an on-board container [24,25]. For light vehicles and passenger cars running under lean conditions, the NH_3-SCR technology is not appropiate because of the volume of the needed ammonia container. Thus, other technologies are being developped, including the SCR with the presence of reductants in the exhaust, e.g. hydrocarbon [26], and the NO_x storage and reduction (NSR) [27-34], which seems to be the most promising technology and to which is dedicated in this chapter.

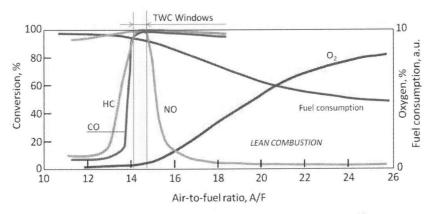

Figure 1. Fuel consumption and TWC behaviour of stoichiometric petrol engines, related to the air-to-fuel ratio.

2.3. General Aspects of the Nsr (No$_x$ Storage and Reduction) Catalysis

Up to day, the NSR is considered as the most promising technology for NO$_x$ removal from diesel engine exhaust gases. The corresponding devices are also denominated lean NO$_x$ traps (LNT). Recent excellent revisions can be found in the literature on this technology [34-36]. Following is a brief summary of the chemical principles used in NSR as to facilitate understanding of next sections.

The NSR catalysts run cyclically under lean environment (oxidizing) and rich environment (reducing), being defined by the corresponding A/F ratios. The concept was introduced by Toyota in the middle 90s [27,33]. While running on the road, lean and rich conditions have to be used in an alternative way [37,38]. Under lean conditions, with excess of oxygen (high A/F), NO$_x$ are adsorbed on the catalyst, and then under rich conditions (A/F<14.63) the stored NO$_x$ are released and reduced. Consequently, an NSR catalyst needs sites for NO$_x$ adsorption (alkaline or earth-alkaline compounds) and also sites for NO$_x$ oxidation and/or reduction (noble metals, as in the TWC technology). Most studies in the literature have used storage materials based on Ba. Also other metals such as Na, K, Mg, Sr and Ca have been used. Thermodynamic and kinetic data demonstrated that basicity of alkaline and earth-alkaline metals is related directly to the NO$_x$ storage capacity, i.e. the storage behavior at 350 °C decreases as follows: K > Ba > Sr ≥ Na > Ca > Li ≥ Mg [34].

The noble metals are normally incorporated with very low percentage, 1-2 wt %. As in the TWC technology, platinum, palladium and rhodium are mostly used [39]. The metal participates into two important steps of the NSR mechanism, the oxidation of NO to NO_2 during the lean period and the reduction of NO_x released during the rich period. In general, it is stablished in the literature that Pt is a good catalyst for NO oxidation, while Rh is more active for NO_x reduction. Obviously, the storage compounds as well as the noble metals should be dispersed on porous materials with high surface area (Al_2O_3, ZrO_2, CeO_2, MgO) washcoated over a monolithic structure, usually cordierite. The most studied formulation in the literature has been Pt-Ba/Al_2O_3, which has also been chosen for this study.

Presently, it is well assumed that the NSR mechanism can be explained by the five following steps, as represented in the upper scheme of Fig. 2 [34,35]:

a. Oxidation of NO to NO_2 (lean conditions, oxidizing environment).
b. Adsorption of NO_x as nitrites or nitrates on the storage sites (lean period, oxidizing environment).
c. Injection and evolution of the used reductant agent (H_2, CO or HC).
d. Release of the stored NO_x from the catalyst surface to the gas stream (rich period, reducing environment).
e. Reduction of NO_x to N_2 (rich period, reducing environment).

The typical NO_x storage and reduction behaviour can be observed in the bottom graph of Fig. 2. At the beginning of the lean period nearly all the NO_x (NO+NO_2) entering the trap is adsorbed, afterwards the NO_x outlet concentration progressively increases due to the successive saturation of the available trapping sites. When saturation is completed, NO_x outlet concentration equals the NO_x inlet concentration. During the subsequent rich period, when H_2 is injected, the adsorbed NO_x species on the catalyst surface react with hydrogen to form N_2O, NH_3 or N_2, resulting in the regeneration of the trap which is again ready for the following lean period.

3. PREPARATION PROCEDURE OF MONOLITHIC NSR CATALYSTS

Most work dealing with NO_x storage and reduction technology have normally used powder catalyst to carry out different studies. However, for real application, NSR catalysts have to be synthesized in a monolithic structure in order to minimize the pressure drop in the catalytic converter [40-42]. The preparation procedure of powder or monolithic catalysts differs notably. While conventional techniques are used for the incorporation of the active phases in powder catalysts, such as wetness impregnation [43-45], the synthesis of monolithic catalysts requires more sophisticated techniques. This section will be focused on the preparation procedure of monolithic NSR catalysts, paying special attention on their final physico-chemical characteristics (dispersion and distribution of the active phases) and their correlation with the activity for NO_x storage and reduction.

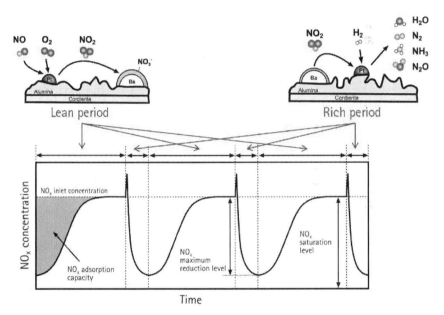

Figure 2. Storage and reduction of NO_x. (a) Schematics of the mechanism; (b) NO_x concentration curves at the exit, during lean and rich periods.

In real application, the mechanical properties of the catalyst are crucial due to the dramatic temperature changes and vibrational strengths that are expected. In this sense, cordierite ($2MgO.2Al_2O_3.5SiO_2$) has been chosen as the base material in automotive application due to its high thermal stability and low expansion coefficient. However, this material exhibits a low surface area which is not suitable for the subsequent incorporation of the active phases. Consequently, the first step of the catalyst preparation consists on the monolithic substrate washcoating with a high surface area oxide, usually alumina.

The most common washcoating procedure is carried out by dipping the monolith into slurry which contains the alumina for washcoating. The monolith is immersed in the slurry for a few seconds and then removed, and the excess of liquid remaining in the channels is blown out with compressed air. This procedure is repeated until the desired Al_2O_3 weight is incorporated as washcoat. It has been previously reported that the characteristics of the final coated monoliths are governed by the properties of the slurry [42], considering as main variables the Al_2O_3 particle size, the Al_2O_3 wt % in the slurry and the pH. Agrafiotis et al. [46] found a threshold value of particle size around 5 μm which is coincident with the size of the cordierite macropores; larger alumina particles do not penetrate into the macropores of the substrate resulting in a poor anchoring of the alumina layer. Therefore, the smaller the particle size in the slurry, the higher the alumina layer anchoring is. In fact, in our previous work [47], the immersion of the monoliths in an alumina slurry with a particle size distribution centered in 1 μm, led to a highly adhered alumina layer, with a weight loss smaller than 0.25 % after the washcoated monolith was immersed in ultrasound bath for 15 min. Regarding the Al_2O_3 wt % in the slurry, two contradictory effects are observed. On one hand, as the slurry concentrates in Al_2O_3, few immersions are required to achieve a given amount of washcoated alumina, but on the other hand, the increase in the slurry viscosity resulted in non-homogeneous coating. The influence of the slurry viscosity on the alumina layer homogeneity has been associated with the ease for the suspension excess to be blown out from the monolith channels [48-50]. Another characteristic to be controlled is the stabilization of the alumina slurry so as to avoid the particles from settling down. Nijhuis et al. [41] suggested that addition of some acid to shift pH between 3 and 4 improved the slurry stabilization. Furthermore, the addition of acetic acid up to 2.5 mol l^{-1} (pH=2.6) decreased considerably the viscosity of the slurry, permitting the use of concentrated Al_2O_3 slurries without penalization in the layer homogeneity [47].

Fig. 3 shows the characterization of a washcoated monolith with scanning electron microscopy (SEM). The washcoating procedure was carried out by immersion of the monolith into alumina slurry with the following characteristics: 20 wt % Al_2O_3, mean diameter of 1 μm and 2.5 mol l^{-1} of acetic acid. Eight immersions of the monolith were needed to achieve around 400 mg of Al_2O_3 over the monolithic substrate (D=L=2 cm). Fig. 3a shows a lower magnification image where the intersection of different channels of the monolith can be observed. As it can be clearly noticed, the original structure of the cordierite was completely covered with alumina. The deposition is preferential in the corners of the channels whereas far away from this position, the alumina layer has a constant thickness of 5 μm. Fig. 3b shows a higher magnification image where a crack in the alumina layer is observed. This image also confirms that the alumina layer was composed of particles around 1 µm in size.

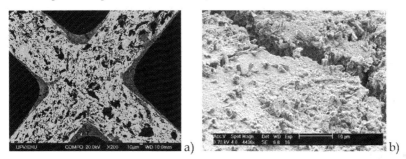

Figure 3. Scanning electron microscoy (SEM) images of the washcoated monolith. (a) Cross section. (b) Higher magnification image of surface in the monolith corner.

The next step in the catalyst preparation is the incorporation of the active phases. As already mentioned, NSR catalysts are usually composed of an alkali or alkali-earth oxide and a noble metal deposited onto the alumina. The most common metal used for NSR catalyst formulation is Pt, whereas BaO is normally used as the storage component [34-36]. The order of the incorporation steps of the active phases Pt and Ba is crucial, especially when operating at higher temperatures; a higher storage capacity is obtained when impregnating Pt/Al_2O_3 with Ba than when impregnating Ba/Al_2O_3 with Pt, increasing the storage value as much as 54 % when adding Ba in the last step [51].

Platinum was incorporated following two different procedures, conventional wetness impregnation (1 monolith) and adsorption from solution (3 monoliths). For the conventional procedure, the channels of the monolith were filled with an aqueous solution containing the desired amount of Pt, using $Pt(NH_3)_4(NO_3)_2$ as a precursor [52]. Then, liquid was evaporated at 80 °C and finally the monolith was calcined at 500 ºC for 4 h. On the other hand, in the adsorption from solution procedure, the monoliths were immersed in an aqueous solution with the adequate concentration of Pt. The pH of the solution was turned basic (11.9) in order to generate an electrostatic attraction between the alumina surface, positively charged, and the Pt precursor, negatively charged $Pt(NH_3)_4^{2+}$ [53-55]. The monoliths were maintained immersed in the solution for 24 h so as to reach the adsorption equilibrium. Then, the monoliths were removed from the solution, the excess of liquid blown out and finally the monoliths were calcined at 450, 500 and 550 ºC, respectively. The four prepared monolith catalysts were tested for their performance in the NSR process.

Irrespective of the calcination temperature, the monolith prepared by wetness impregnation showed the lowest dispersion of platinum (15 %). In the rest of the samples, platinum dispersion decreased as the calcination temperature increased, from 54 % at 450 ºC to 46 % at 500 ºC and finally to 19 % at 550 ºC. Then, 500 ºC was chosen as the optimal calcination temperature as a good compromise between platinum dispersion and thermal stabilization of the catalyst. Fig. 4b shows the platinum particle size distribution determined from the transmission electron microscopy image (Fig. 4a) for a Pt/Al_2O_3 sample prepared by adsorption from solution and calcined at 500 ºC. As it can be observed, the Pt particles are fairly dispersed over the alumina washcoat with a mean particle size of 1.3 nm.

The last step in the NSR catalyst preparation is the incorporation of the NO_x storage component, i.e. barium. The precursor used was barium acetate [52] and two different procedures were followed: wetness impregnation and incipient wetness impregnation (also known as dry impregnation). For wetness impregnation, the monolith channels were filled with an aqueous solution containing the desired amount of barium. Then, the monolith was dried and calcined. Alternatively, for incipient wetness impregnation, the monolith was immersed in an aqueous solution with an adequate concentration of barium acetate for a few seconds; then, the monolith was removed and the liquid in the channels was blown out with compressed air. Thus only the liquid retained in the pores of the alumina remained in the monolith. In order to determine the distribution of Ba in the catalyst, the monolith was divided into 8 pieces and the content of barium was determined by inductively coupled plasma mass spectrosmetry (ICP-MS). It was found that the distribution of barium resulted in an egg-shell type for wetness impregnation, whereas the incorporation of Ba by incipient wetness impregnation led to more homogenous distribution. Table 3 resumes the preparation procedure, the catalyst physico-chemical characteristics and the NO_x storage achieved with the prepared catalysts (A, B, C and D).

The activity of the prepared catalysts was tested in a vertical downstream reactor with a feedstream composed of 380 ppm NO and 6 % O_2 during the lean period (150 s) and 380 ppm NO and 2.3 % H_2 during the rich period (20 s) using nitrogen as the balance gas in both cases. The total flowrate was 3365 ml min^{-1} that corresponds to a gas hourly space velocity (GHSV) of 32,100 h-1. Fig. 5 shows the NO_x concentration profile at the reactor exit for A, B, C and D catalysts. As it can be observed, the NO_x concentration profile is always below the inlet value (380 ppm NO) which evidences the activity of the prepared catalysts for the storage of NO_x. During the storage-reduction cycles, the typical NO_x concentration profile was recorded [56,57]. At the beginning of the lean period practically all the NO_x is stored, and consequently its concentration at the reactor exit is very low. Then, as the lean period time increases, the storage sites become saturated and the NO_x concentration at the reactor exit gradually increases. During the rich period, the NO_x stored are released and reduced with the injected hydrogen, leaving the catalyst surface clean for the subsequent storage period.

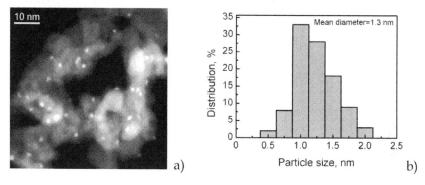

Figure 4. Platinum dispersion measurements. a) Transmision electron microscopy image. b) Platinum particle size distribution.

Table 3. Preparation procedure, physico-chemical characteristics and NO_x storage capacity for the prepared catalysts A, B, C and D. WI: Wetness impregnation; ADS.: Adsorption from solution; DI: Dry impregnation. The dispersion values are estimated based on powder Pt/Al_2O_3 samples. * Distribution: (+) not good (++) good (+++) very good.

	A	B	C	D
Pt incorporation	WI	ADS	ADS	ADS
Calcination T, °C	500	550	500	500
Dispersion,%	15	19	46	46
*Distribution	+	+++	+++	+++
% Pt	0.72	1.43	1.34	1.14
Ba incorporation	WI	WI	WI	DI
*Distribution	+	+	+	++
% BaO	13.1	13.3	16.3	25.2
NO_x storage capacity, %	47.5	55.7	69.6	76.7

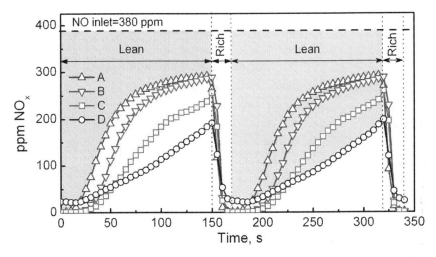

Figure 5. NO$_x$ concentration profile at the reactor exit for two consecutive NO$_x$ storage and reduction cycles for catalysts A, B, C and D.

The NO$_x$ storage capacity is related with the area between the NO inlet level and the NO$_x$ outlet concentration profile; the lower NO$_x$ concentration at the reactor outlet the higher activity of the catalyst for the storage of NO$_x$. Thus, among the prepared catalysts, catalyst D was found to be the most active (gray area on the rigth graph) and catalyst A (blue area on the left graph). Quantification of the NO$_x$ storage capacity can be found in Table 3 with the following order from the least to the most active: A<B<C<D, according to the physico-chemical characteristics of the samples. It is well known that the Pt-Ba pair is the responsible for the storage of NO$_x$ and that proximity between both metals is beneficial for the process [47,58,59]. Consequently, catalyst A resulted in the less active sample due to the low platinum dispersion and non-homogeneous distribution of both Pt and Ba, as it was prepared by conventional wetness impregnation (WI) of Pt and Ba. For catalyst B, the incorporation of Pt by adsorption from solution (ADS) increased Pt dispersion, but just slightly as the higher calcination temperature (550 ºC) also provokes some platinum sintering. Higher Pt loading and dispersion were identified as responsible for better storage capacity, from catalyst A (47.5 %) to catalyst B (55.7 %). The lower calcination temperature (500 ºC) for catalyst C provided much higher dispersion, thus enhancing the NO$_x$ storage capacity up to 69.6 %. Furthermore, the incorporation of Ba by dry impregnation (DI) in catalyst D provided a better distribution of barium over the monolith and consequently increased the Pt-Ba proximity resulting in the best storage capacity (76.7 %), i.e. 76.7 % of NO at the inlet was trapped in the catalyst.

4. THE CHEMISTRY OF NO$_x$ STORAGE AND REDUCTION

Many studies are available in the technical literature dealing with the storage step of NSR catalysts. Particularly relevant in this field is research by Forzatti et al. [60-65] and Fridell et al. [66-68]. In situ FTIR spectroscopy has been found to be a very useful tool and several studies have been made on adsorbed NO$_x$ species, though the assignment of peaks is still under debate. Takahashi et al. [27] were pioneers in studying the interaction of NO$_x$ over NSR catalysts and assigned the 1350 cm^{-1} peak to the nitrate anion. Fig. 6a shows the FTIR spectra of Pt-Ba/Al$_2$O$_3$ sample after it was exposed during 20 minutes to a feedstream composed of 440 ppm NO and 7 % O$_2$ using N$_2$ as the balance gas. As it can be observed, the FTIR spectra changed very significantly with the operating temperature. Bridged nitrites situated at 1220 cm^{-1} [66] were dominant when the adsorption was carried out below 250 ºC, whereas the dominant species became ionic nitrates (asymmetric and symmetric modes of monodentate nitrates) located at 1332 and 1414 cm^{-1} [69] for temperatures above 250 ºC. This shift in the adsorption mode from nitrites to nitrates with temperature had been already reported in the literature [62,63,66,67,70].

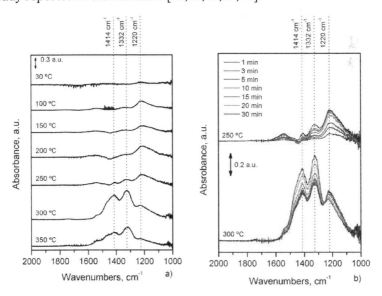

Figure 6. FTIR experiments with powder Pt-BaO/Al$_2$O$_3$. (a) Absorbance signals at different temperatures and after 20 min of contact time. (b) Absorbance of the sample at 250 and 300 ºC with increasing contact time.

All spectra included in Fig. 6a were recorded after the sample had been exposed to the lean gas mixture for 20 min. However, it can be interesting to examine the evolution of adsorbed species with increasing contact time. Owing to the fact that the adsorption mode of NO_x changed from nitrites at 250 ºC to nitrates at 300 ºC, several FTIR spectra were recorded at different contact times for those temperatures. The first spectrum was recorded after the sample had been exposed to the lean gas mixture for 1 min while the last one was taken after 30 min. As revealed by Fig. 6b, when the adsorption was carried out at 250 ºC nitrite was immediately formed upon admission of NO, whereas nitrate formation was delayed. Furthermore, the intensity of the peaks corresponding to nitrite (1220 cm^{-1}) and nitrate (1322, 1414 cm^{-1}) increased nearly in the same extent with increasing contact time, which means that there was no conversion from nitrites to nitrates or that conversion from nitrites to nitrates and formation of additional nitrite species occurred simultaneously. In short, below 250 ºC nitrite was the dominant adsorption species even at contact times of 30 min, which is much longer than in real operation (1-2 min).

On the other hand, the adsorption pattern resulted completely different at 300 ºC (Fig. 6b). From the beginning of the adsorption, the intensity of the bands assigned to ionic nitrates was higher than nitrite. Moreover, it can be noticed that the adsorption peak assigned to nitrites resulted maximum in the first minute of storage and then gradually decreased till minimum after 30 min of contact time. Thus, it can be concluded that there is a shift from nitrite to nitrate when increasing contact time which can be associated with the oxidation of nitrites to nitrates under the lean gas mixture.

In early ages of NSR catalysts, Fridell et al. [66] proposed a three step mechanism in which NO_2 is at first loosely adsorbed on BaO as a BaO-NO_2 species; this species then decomposes to BaO_2 and NO (which is released in the gas phase) and finally barium peroxide reacts with the gas-phase NO_2 to give barium nitrate which can be illustrated as:

$$NO_2 + BaO \rightarrow BaO\text{-}NO_2 \tag{1}$$

$$BaO\text{-}NO_2 \rightarrow BaO_2 + NO \tag{2}$$

$$2NO_2 + BaO_2 \rightarrow Ba(NO_3)_2 \tag{3}$$

The overall stoichiometry of NO_2 adsorption implies the release of one molecule of NO for the consumption of three molecules of NO_2. This reaction is known as the NO_2 disproportionation and has been widely reported for NSR catalysts [62,71-74]:

$$3NO_2 + BaO \rightarrow Ba(NO_3)_2 + NO \quad (4)$$

The formation of nitrate species following the reactions above described, clearly evidences that the oxidation of NO to NO_2 is a preliminar and necessary step for the adsorption of NO. The reaction mechanism used to describe the NO oxidation consists of the following adsorption and desorption steps [75,76]:

$$O_2 + 2Pt \rightarrow 2Pt\text{-}O \quad (5)$$

$$NO + Pt \rightarrow NO\text{-}Pt \quad (6)$$

$$NO + Pt\text{-}O \rightarrow NO_2\text{-}Pt \quad (7)$$

$$NO\text{-}Pt + Pt\text{-}O \rightarrow NO_2\text{-}Pt + Pt \quad (8)$$

$$NO_2\text{-}Pt \rightarrow NO_2 + Pt \quad (9)$$

On the other hand, FTIR experiments showed that apart from nitrates, surface nitrites are also formed during adsorption of NO over the Pt-Ba/Al_2O_3 catalyst. It has been proposed that barium peroxide formed in reaction (2) could also react with NO to form nitrites [62,71,77]:

$$BaO_2 + 2NO \rightarrow Ba(NO_2)_2 \quad (10)$$

BaO_2 can also be formed by an alternative route to reactions (1) and (2), in which NO_2 is not involved, as the following

$$O\text{-}Pt + BaO \rightarrow BaO_2 + Pt \quad (11)$$

The close proximity of BaO to Pt sites promotes spillover of the oxygen adatoms from Pt to BaO. From FTIR spectra shown in Fig. 6a it can be deduced that, at 300 ºC, Pt catalyzes the formation of barium nitrate species from nitrite species, which is illustrated as:

$$Ba(NO_2)_2 + 2O\text{-}Pt \rightarrow Ba(NO_3)_2 + 2Pt \quad (12)$$

Thus, from the experiments shown in Fig. 6, two parallel routes can be described for the adsorption of NO_x, which are in concordance with the most accepted mechanism by Forzatti et al. [65]. The first route is called "nitrite route" where NO is oxidized at Pt sites and directly stored onto Ba neighbouring sites in the form of nitrite ad-species (reaction 10), which can be progressively transformed into nitrates depending on reaction temperature (reaction 12). The second route is called "nitrate route" which implies the oxidation of NO to NO_2 on Pt sites, followed by NO_2 desproportionation on Ba sites to form nitrates with the giving off NO into the gas phase (reaction 4).

The regeneration step of NSR catalysts is not so well understood as the storage step. Several studies have been published on the chemistry and mechanisms that rule the reduction of NO_x ad-species by H_2. The nitrite and nitrate decomposition can be driven by either the heat generated from the reducing switch [78,79], or the decrease in oxygen concentration that lowers the equilibrium stability of nitrates [34,80]. However, under near isothermal conditions, it has been found that the reduction process is not initiated by the thermal decomposition of the stored nitrates, but rather by a catalytic pathway involving Pt [45]. The reduction of stored nitrates and nitrites leads to the formation of different nitrogen containing species, such as N_2, NH_3 and N_2O along with H_2O. The objective of the NSR operation is to maximize the conversion of NO into N_2, avoiding the formation of NH_3 and N_2O as far as possible. The operational conditions to run efficiently the NSR process are discussed in detail in section 5.

Fig. 7 shows the concentration profiles of NO, NO_2, NH_3, N_2O and H_2O, and the evolution of the MS-signal for N_2, O_2 and H_2 during the regeneration step when the reaction was carried out at 330 ºC. The feedstream composition during the lean period was 975 ppm NO, 6 % O_2 and Ar to balance, extending the length of this period until complete saturation of the catalyst was obtained. Afterwards, during the rich period, oxygen was replaced by 0.6 % H_2 for 500 s. As can be observed in Fig. 7, before the regeneration period started the sum of NO and NO_2 concentration was close to the inlet value (975 ppm), confirming that the catalyst was saturated. The presence of NO_2 at the reactor exit is due to the oxidation of NO by Pt sites as described in eqns. (5)-(9). When the rich feedstream contacts the catalyst (t=0), the NO and NO_2 concentrations are progressively reduced, eventually reaching 0 ppm. At the very beginning of the rich period a sudden increase in the NO and NO_2 concentrations can be observed due to the release of adsorbed NO_x as a consequence of the decrease in oxygen partial pressure that reduces the stability of the stored nitrates and nitrites. Meanwhile, the incoming H_2 reacts with adsorbed NO_x to form N_2, NH_3 and N_2O. As can be observed in Fig. 7, the formation of N_2 and N_2O is detected immediately after the reduction period started, whereas the detection of H_2O and NH_3 was delayed, the later in a much more extent. On the other hand, the complete consumption of H_2 during the initial period of the regeneration together with the rectangular shape of the H_2O and N_2 formation curve indicates a "plug-flow" type of the regeneration mechanism. As several authors have already reported [45,78,81,82], the hydrogen front travels through the catalyst bed with complete regeneration of the trapping sites as it propagates down the bed with regeneration time. After the required time, complete regeneration of the trap is obtained, i.e. no nitrates or nitrites are present in the catalyst surface, and consequently H_2 is detected at the reactor outlet.

The reactants and product profiles shown in Fig. 7 are in agreement with mechanistic aspects of the regeneration already reported [45,81,83,84]. The reduction of stored nitrates with hydrogen has been found to occur by the following reactions:

$$Ba(NO_3)_2 + 8H_2 \rightarrow 2NH_3 + BaO + 5H_2O \quad (13)$$

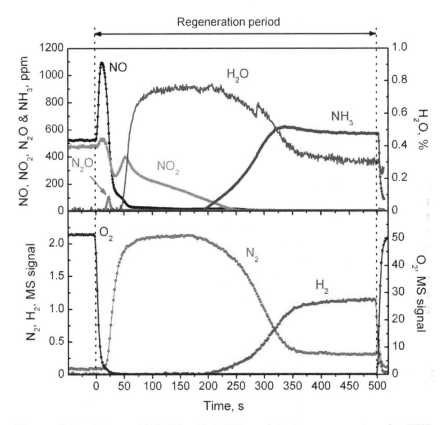

Figure 7. Evolution of NO, NO_2, N_2O, NH_3 and H_2O concentrations by FTIR and MS signals of O_2, N_2 and H_2, during Pt-BaO/Al_2O_3 catalyst regeneration at 330 ºC.

$$Ba(NO_3)_2 + 5H_2 \rightarrow N_2 + BaO + 5H_2O \qquad (14)$$

Lietti et al. [84] reported that during reduction of stored nitrates at 100 ºC, reaction (13) accounted for almost all the H_2 consumption, demonstrating that stored nitrates were reduced efficiently and selectively (>90 %) to ammonia. On increasing the reduction temperature, nitrogen formation was promoted due to reaction (15) where the formed ammonia continued to react further with stored nitrates to form nitrogen.

$$Ba(NO_3)_2 + 10NH_3 \rightarrow 8N_2 + 3BaO + 15H_2O \tag{15}$$

Thus, nitrogen formation involves a two-step pathway: the fast formation of ammonia by reaction of nitrates with H_2 (reaction 13) and the subsequent conversion of the ammonia formed with stored nitrates leading to the selective formation of N_2 (reaction 15). This overall mechanism for nitrates reduction during LNT regeneration has been confirmed by Pereda-Ayo [85] using isotope labelling techniques and explains the evolution of products at the reactor exit shown in Fig. 7. When the hydrogen front enters the catalyst, the stored NO_x are thought to be converted mainly to ammonia, with total H_2 consumption. Then, the ammonia formed in the regeneration front reacts further with stored nitrates located downstream to give nitrogen. Thus, N_2 is detected as soon as the regeneration period starts, but no ammonia can be detected since it was completely consumed. As the regeneration time increases and the hydrogen front moves forward, the ammonia formed has fewer nitrates to react with, and therefore some NH_3 starts to leave the catalyst unreacted, being detected at the reactor outlet.

5. ANALYSIS OF ENGINEERING VARIABLES OF THE NSR PROCESS

The importance of the catalyst properties, chemical composition, structure, morphology and, in special dispersion and distribution of the metallic phases on its behavior in storing and reducing NO_x has been well reviewed in the work of Roy and Baiker [35]. Generally, research in the scientific open literature has studied independently the two stages, storage and reduction, to advance in the understanding of the mechanisms that happen in each stage. Up to now many papers on the stage of storage have been published [e.g. 60, 63, 66-68, 70, 72, 87], but notably less on the stage of regeneration (liberation and reduction) [e.g. 39, 81, 86, 88, 89]. However, very few studies have considered the whole operation, where the lean and rich periods occur successively as in the real application.

In fact, still scarce relations have been proposed between storage, regeneration and product distribution as determined in laboratory, and the optimization of the conditions in which the catalytic converter should operate in automobiles. It must be mentioned that engineering parameters, such as gas hourly spatial velocity (GHSV), residence time in the converter, and the lasting time of the periods of storage and regeneration, influence significantly the behavior of lean NO_x traps (LNTs). Kabin et al. [90] studied the storage and reduction of NO_x on model monolithic Pt-BaO/Al_2O_3 catalysts to relate the percentage of the NO_x trapped during storage (trapping efficiency) and the reduced NO_x percentage (average NO_x conversion) with the load of Ba (6-25 %) and the GHSV (30,000-120,000 h^{-1}). These authors concluded that the dependence of trapping efficiency on the storage period duration provides a good estimation of the time needed to get a given average conversion during the whole NSR process. More recently, Clayton et al. [91] determined the effects of the catalyst temperature, the composition of the rich stream (NO, H_2, O_2), duration of lean and rich periods and the H_2/NO ratio on the average conversion and product selectivity of a commercial Pt-BaO/Al_2O_3 catalyst. The NO_x average conversion resulted maximum at 300 ºC, and also the trapping efficiency resulted maximum at the same temperature. The selectivity to N_2 exhibited a maximum at slightly superior temperature, where the selectivity to NH_3 was minimized.

Thus, to define the optimal operation in the NSR technology, the efficiency of both the storage and reduction steps but also the global NSR efficiency must be studied. This section is devoted to set a definition for the global NSR efficiency of the process, obviously related to the catalyst behavior, the NO_x storage and reduction mechanisms, and the kinetics of the release and reduction of NO_x during the regeneration phase. The defined parameter must account the byproduct formation to maximize the NO_x reduction efficiency towards N_2. The process selectivity depends notably on the lean and rich period duration. For this purpose, experimental runs have been carried out with storage period duration in the order of minutes, followed by rich injection periods during some seconds. The effect of the lean and rich period duration and the reductant concentration should be analyzed, stating as objective functions for optimization the storage capacity in the lean period, the NO_x conversion in the rich period, and the selectivity towards N_2O/NH_3/N_2. Also a bidimensional analysis of operational variables, including temperature and hydrogen concentration, will be made in this section.

The experiments were carried out with a homemade 1.2 % Pt-15 % BaO/Al_2O_3 monolith catalyst, prepared as explained in section 3 [1,47]. Each time some operational variable was altered, at least ten successive lean-rich cycles were proceeded in order to assure a new stable state of the system, and then the performance was monitored.

5.1. Definition of Response Parameters

To evaluate the performance of the catalyst during the lean and rich periods the NO_x storage capacity, NO_x conversion and N_2/NH_3 selectivities have been determined from the concentration curves at the reactor exit monitored during the experimental storage-reduction cycles.

The total NO_x stored during the lean period, was calculated as

$$NO_x^{stored}(\mu mol\ NO) = (NO^{in})_L - (NO_x^{out})_L \qquad (16)$$

where $(NO^{in})_L$ is the total amount of NO fed during the lean period and $(NO_x^{out})_L$ is the total amount of NO and NO_2 leaving the reactor during the same period. These amounts correspond to the areas graphically represented in Fig. 8, which can be calculated by the corresponding numerical integrations.

When the cumulative NO_x trapped (eqn. 16) is expressed as a percentage of the NO_x fed, then it is referred to as the NO_x storage capacity,

$$\mu_{STO} = \frac{NO_x^{stored}}{(NO^{in})_L} \times 100 \qquad (17)$$

The catalyst performance during the rich period was described, on the one hand, by the reduction conversion (X_R) defined as the percentage of NO_x reduced over the total amount of NO_x to be reduced. The latter accounts for the sum of the NO_x stored during the lean period plus the NO continuously fed during the rich period. Then,

$$X_R(\%) = \frac{NO_x^{reduced}}{NO_x^{to\ be\ reduced}} \times 100 = \frac{\left[NO_x^{stored} + (NO^{in})_R\right] - (NO_x^{out})_R}{\left[NO_x^{stored} + (NO^{in})_R\right]} \times 100 \qquad (18)$$

On the other hand, the NO_x reduction conversion needs to be complemented with the selectivity to different nitrogen species, including dinitrogen oxide, ammonia and nitrogen. Then, selectivities were defined as

$$S_{NH_3} = \frac{NH_3^{out}}{NH_3^{out} + 2N_2^{out} + 2N_2O^{out}} \times 100 \qquad (19)$$

$$S_{N_2O} = \frac{2N_2O^{out}}{NH_3^{out} + 2N_2^{out} + 2N_2O^{out}} \times 100 \qquad (20)$$

Fig. 8 shows over a model profile for NO_x and NH_3 molar flows at the exit of the reactor the areas corresponding to the component amounts leaving the reactor during the regeneration period. The N_2O amount leaving the reactor is calculated similarly from the corresponding outlet profile. The FTIR technique is not able to analyze N_2 so that the amount of this component should be calculated from the nitrogen mole balance

$$NO_x^{stored} + (NO^{in})_R = NH_3^{out} + 2N_2^{out} + 2N_2O^{out} + (NO_x^{out})_R \qquad (21)$$

i.e. the amount of NO_x stored plus the NO_x amount fed during the lean period equals the NO_x amount leaving the reactor and those amounts converted into NH_3, N_2 and N_2O. Then, the selectivity to N_2 can be expressed as

$$S_{N_2} = \frac{2N_2}{NH_3^{out} + 2N_2^{out} + 2N_2O^{out}} \times 100 = \frac{\left[NO_x^{stored} + (NO^{in})_R\right] - \left[NH_3^{out} + 2N_2O^{out} + (NO_x^{out})_R\right]}{NH_3^{out} + 2N_2^{out} + 2N_2O^{out}} \times 100 \qquad (22)$$

On the other hand, the nitrogen mole balance can be checked if one can determine the amount of N_2 at the reactor exit by the adequate analysis technique, e.g. quantitative mass spectrometry.

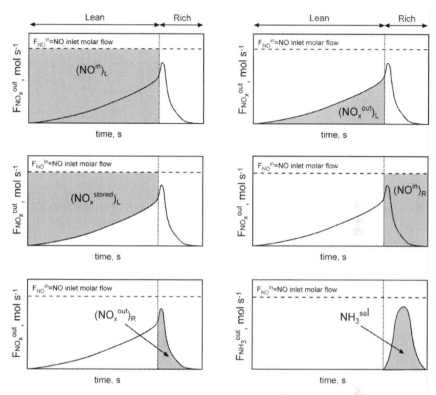

Figure 8. Graphical representation of areas corresponding to magnitudes needed to define parameters for evaluation of the catalyst performance in the NSR process.

The parameters above defined are useful to compare independently the NO_x storage capacity during the lean period and the NO_x reduction conversion during the rich period. The N2/NH3 selectivities have been calculated averaged over the whole cycle, as peaks corresponding to those compounds can be seen at the outlet during the rich time but also continuing during the subsequent lean period (seeFig. 9b later). In the case of a conventional $Pt-BaO/Al_2O_3$ NSR system, the catalyst should operate to exhibit high NO_x storage capacity with also high selectivity to N_2. Thus, definition of a single parameter giving information of the trap performance over the whole storage-reduction cycle would be very convenient to know how efficiently the NSR system is running. This global parameter should take into account the storage capacity, the reduction conversion and the selectivity of the reaction, giving a general vision of the efficiency of the whole NSR process. Thus, the global NSR efficiency, referred to the N2 production over the total amount of NO_x fed, can be calculated as

$$\varepsilon_{NSR} = \frac{2N_2^{out}}{(NO^{in})_L + (NO^{in})_R} \times 100 = \frac{2N_2^{out}}{NH_3^{out} + 2N_2^{out} + 2N_2O^{out} + (NO_x^{out})_R} \times 100 \quad (23)$$

5.2. Optimal Control of the Nsr Technology by Managing the Amount Of Reductant Injected During the Regeneration Period

NO_x storage-reduction experiments were carried out in a downflow steel reactor. The monolithic catalyst (25 mm in length and diameter, 3.5 g) was placed in the bottom part of the reactor and the set was introduced in a 3-zone oven. The temperature at the entry and exit of the reactor was continuously monitored. The experimental conditions are shown in Table 4. The feedstream during storage was 380 ppm NO/6 % O_2/N_2. Gases were fed through mass controllers with a total volumetric flow of 3,365 l min⁻¹, corresponding to a GHSV of 32,000 h⁻¹ (STP).

The problem was stated as follows: to find the values of the operational variables, including lasting time of the storage period (lean mixture), lasting time of the regeneration period (rich mixture) and hydrogen concentration injected during the regeneration period, which allow to achieve the maximum NSR efficiency (Eq. 23).

Table 4. Experimental conditions.

Operational parameters	Values
Temperature, °C	330
Total volumetric flow, l min⁻¹ (STP)	3.365
Spatial velocity, GHSV, h⁻¹	32,100
Lean period time duration, s	145, 290, 595
Rich period time duration, s	16 - 47
Lean mixture composition	380 ppm NO, 6% O_2, N_2 to balance
Rich mixture composition	380 ppm NO, 0.41 – 2.36% H_2, N_2 to balance

5.2.1. Effect of the h_2 concentration in the regeneration stream on no_x storage and reduction

Fig. 9 shows the effect of hydrogen concentration in the rich stream on the NO_x and NH_3 concentrations at the exit of the reactor, for experiments carried out at 330 ºC. During the lean period the stream composition was 6 % O_2 and 380 ppm NO, with N_2 to balance. After 145 s of lean period the oxygen was shifted to hydrogen at different concentrations (0.79, 1.1 and 2.32 %), maintaining 380 ppm NO in the feedstream for the regeneration period of 25 s.

The evolution of NO_x concentration at the reactor exit during the storage and regeneration periods is that typical for the NSR process (see Fig. 2), i.e. at the beginning of the lean period all amount of NO_x is stored, and this amount is gradually reduced as the adsorption sites are being saturated, then increasing the NO_x exiting the reactor. It can be seen in Fig. 9a, nevertheless, that the NO_x concentration at the end of the lean period (145 s) did not reach the initial concentration (380 ppm NO), i.e. the catalyst was not completely saturated. The monitored values resulted in 240, 170 and 145 ppm NO_x for runs 1, 2 and 3, respectively.

After 145 s of lean period, the shifting of oxygen by hydrogen to the entry of the reactor provokes the release of the previously stored NO_x which is eventually reduced to N_2O, NH_3 and N_2, according to the mechanisms explained in section 4. The evolution of ammonia concentration at the reactor exit is shown in Fig. 9b. At 330ºC, however, N_2O was not practically apreciated at the reactor exit.

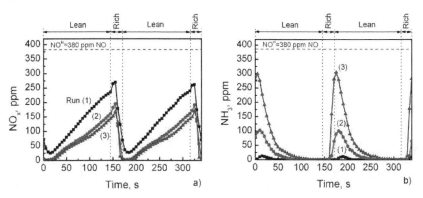

Figure 9. Concentration of (a) NO_x and (b) NH_3 at the reactor exit for two consecutive storage-reduction, for different H_2 concentration during the rich period: (1) 0.79 % H_2; (2) 1.1 % H_2; (3) 2.32 % H_2. t_L=145 s, t_R=25 s.

The hydrogen concentration influences significantly the formation of ammonia at the exit, from 15 ppm NH_3 for 0.79 % H_2 to a maximum of 300 ppm NH_3 for 2.32 % H_2. In fact, the higher H_2 concentration in the reduction stream the higher NH_3 concentration at the exit, as previously reported [56,78,87-89,91]. These experiments also confirmed the delay in the ammonia detection at the reactor exit related to the beginning of the rich period, as it was observed in the experiments of section 3.

The amount of hydrogen supplied during the rich period also influences the storage capacity and distribution of products. In fact, the storage capacity (μ_{STO},eqn.17) increased with H_2 concentration, resulting in 40.8, 77.3 y 80.5 % for runs 1, 2 and 3, respectively. Similarly, selectivity to NH_3 (S_{NH3}, eqn. 19) also increased with H_2 concentration, resulting in 2, 10 and 29 %, respectively. This influence can be better observed in Fig. 10, where values of the response variables corresponding to additional H_2 concentration experiments have been included. The response variables calculated have been: storage capacity (μ_{STO}, eqn. 17), NO_x reduction conversion (X_R, eqn. 18), selectivity to N_2 (S_{N2}, eqn. 22) and NSR efficiency (ε_{NSR},eqn.23).

Fig. 10a evidences a linear increase of the storage capacity (red line) with H_2 concentration up to 1.1 %. Above 1.1 % H_2, the storage capacity was maintained almost constant in about 80 %. This evolution can be explained from the regeneration mechanism proposed in section 4 [84,85]. During the regeneration step there exits a hydrogen front that travels along the catalyst while regenerating the adsorption sites. If the amount of hydrogen fed is enough, the regeneration front will travel until regeneration of the complete trap. On the contrary, if the reductant is in defect, the regeneration front will not arrive to the final part of the reactor, and the adsorption sites downstream could not be regenerated. This explains that 0.79 % H_2 was not able to regenerate the whole catalyst resulting in a limited storage capacity (40.8 %). While increasing the H_2 concentration, the regeneration front was able to reach more advanced positions and thus regenerates more adsorption sites, resulting in higher storage capacity during the lean period. Above 1.1 % H_2 it is assumed that the regeneration front travels the whole catalyst with total regeneration of the trap. Thus, higher H_2 concentration did not produce significant variations of the storage capacity. Another explanation was supplied by Clayton et al. [86] based on the different storage regions related to the barium phase storage sites based on their proximity to Pt crystallites. They associated the extension of these regions with the platinum dispersion and the reaction temperature.

The reduction conversion (blue line) follows a similar trend to the storage capacity (Fig. 10a). Hydrogen concentration above 1.1 % resulted in almost total conversion (97 %), whereas lower concentrations resulted also in lower conversion level. In fact, when the supply of H_2 is not enough to allow the hydrogen front achieving the complete regeneration of the trap, e.g. 0.79 % H_2, the NO_xreduced/NO_x released ratio is increasing, then decreasing the NO_x reduction conversion to 85 %.

Concerning selectivity N_2/NH_3 (remember that N_2O concentration was negligible at 330 ºC), Fig. 10ashows that formation of ammonia decreased with lower hydrogen concentration during the rich period, being very low with 0.79 % H_2. Thus, when the catalyst regeneration was carried out under low hydrogen concentration (0.79 % H_2) the selectivity to N_2 is practically total and only 15 ppm NH_3 were detected. When increasing the H_2 concentration, the selectivity to N_2 decreased progressively in favour of ammonia (Fig. 9b). This agrees with the observation of Clayton et al. [91] that nitrogen selectivity increased with the NO_x/H_2 ratio.

The trends of NO_x storage capacity, NO_x conversion and N_2/NH_3 selectivity above explained, suggest differentiation of two different zones in Fig. 10, limited to each other by 1.1 % H_2. In zone A occurs that hydrogen is the reactant that limits the reduction of NO_x, and complete NO_x reduction cannot be achieved (Fig. 8, 0.79 % H_2). On the contrary, in zone B the reaction occurs with excess of hydrogen and NO_x are completely reduced. This is also in agreement with the fact that in zone A the NO_xstorage capacity is limited as all barium sites cannot be regenerated, whereas in zone B the excess of hydrogen enhances the ammonia formation.

Fig. 10b shows the evaluation of NSR efficiency (εNSRεNSR) with H_2 concentration. As already mentioned, this can be considered as a global parameter that considers the complete storage-reduction cycle as determining the molar amount of nitrogen at the reactor exit over the molar amount of NO at the entry, expressed as percentage. In fact, with some simple mathematical rearrangements of eqns. (17)-(23), the relationship between the NSR efficiency and the previous response variables can be found, which is expressed as:

$$\varepsilon_{NSR} = X_R S_{N_2}(\mu_{STO}\tau_L + \tau_R) \qquad (24)$$

where τLandτRτLandτR are the dimensionless lean and rich times

$$\tau_L = \frac{t_L}{t_L + t_R} \qquad \tau_R = \frac{t_R}{t_L + t_R} \qquad (25)$$

The opposite trend shown by SN2SN2 and μSTOμSTO with the amount of hydrogen fed during the rich period (Fig. 10a), makes the NSR efficency to reach a maximum at some intermediate value of % H$_2$ (eqn. 17), as seen in Fig. 10b. With low % H$_2$ (zone A) high SN$_2$ is achieved but μ$_{STO}$ is limited, whereas with high % H$_2$ (zone B) low SN$_2$ (high formation of NH$_3$) is achieved but μ$_{STO}$ is maintained maximum. Then, the maximum of ε$_{NSR}$ is achieved for 1.1 % H$_2$, just the border between zones A and B, where the amount of hydrogen is that needed to make the complete regeneration of barium sites but not more to avoid formation of ammonia.

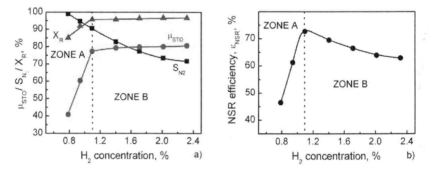

FIGURE 10. a) Evolution of the NO$_x$ storage capacity, reduction conversion and selectivity to nitrogen as a function of H$_2$ concentration during regeneration. (b) NSR efficiency vs. H$_2$ concentration.

5.2.2. Influence of storage and regeneration period duration on the no$_x$storage and reduction

In the previous section all experiments were performed with same duration of the lean and rich periods, i.e. t_L=145 s and t_R=25 s, and varying only the H$_2$ concentration during the rich period. On the other hand, it can be concluded that the total amount of hydrogen fed during the rich period determines the maximum NSR efficiency. Obviously this amount of hydrogen can be considered proportional to the product CH2×tRCH2×tR, so that it can also be varied by modifying the duration of the rich period. Pereda–Ayo et al. [56] made experiments looking for different combinations of pairs (CH2,tR)(CH2,tR)that achieved maximum NSR efficiency, when the lean period duration was maintained in 145 s. The results of these experiments can be represented as the locus of all these combinations as shown in Fig. 11 (t_L=145 s, red curve) and defines the isocurve of operational conditions to carry out the global NSR process efficiently.

The shape of the isocurve represented in Fig. 11 indicates the inverse relationship between the regeneration time and the reductant concentration to achieve an efficient NSR process, i.e. the shorter reduction time, the higher reductant concentration needed to achieve maximum efficiency. This finding implies again that the supply of the reductant H_2 is controlling the NSR process. This was also observed by Mulla et al. [81] that measured the time required for regenerate the trap catalyst by the width of the N_2 pulse in a mass spectrometer. Analogously, Nova et al. [45] had also noticed that the N_2 production during the regeneration of a Pt-BaO/Al_2O_3 catalyst was limited by the amount of H_2 fed to the reactor.

5.2.3. Extension of the duration of lean period on the nsr performance

To further investigate the optimal conditions to operate the NSR process efficiently, the duration of the storage period was varied, extending the lean period duration from 145 to 290 and 595 s [56]. Again, analogous NO_x storage and reduction experiments were performed with those extended times and the same protocol as before. It has been verified that the same storage capacity was obtained for a given lean period duration, resulting in additional isocurves shown in Fig. 11. As expected, the NO_x storage capacity decreased as the lean period duration increased. When the lean period is longer the catalyst is closer to the saturation level, and consequently the NO_x storage capacity decreases, i.e. 77, 55 and 35 % for t_L = 145, 290 and 595 s, respectively. As for selectivity to nitrogen, very similar values were found around 90 %, irrespective of the studied variables (CH2,tR,tL)(CH2,tR,tL), provided that the operation is achieved with maximum efficiency (all points in every isocurve in Fig. 11).

As noted above, the nitrogen production during the regeneration of the catalyst is limited by the amount of hydrogen fed to the reactor. Likewise, increasing the hydrogen supply rate is expected to have a linear effect on the overall rate of NO_x reduction. For our experiments, Fig. 12 shows a linear effect of the hydrogen concentration fed during the rich period on the overall NO_x reduction rate, independent of the duration of the lean period, thus suggesting again that the regeneration step is limited by the amount of hydrogen fed. The linear relationship implies that the time required for complete regeneration should be inversely proportional to the reductant amount of hydrogen fed, as shown in the isocurves of Fig. 11. Mulla et al. [81] also reported the overall rate for NO_x reduction as a linear function of rate of flow of H-atoms in the form of H_2 or NH_3 at 300 ºC. Their observations also confirmed that the regeneration process was not mass transfer or kinetically limited, but it was controlled by the supply of the reductant H_2.

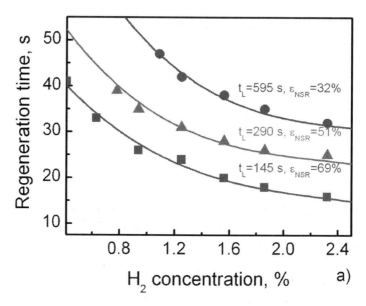

Figure 11. Operation map of the Pt-Ba/Al$_2$O monolithic catalyst. Relationship between operational variables (CH2,tP,tR)(CH2,tP,tR) for carrying out the NSR process efficiently.

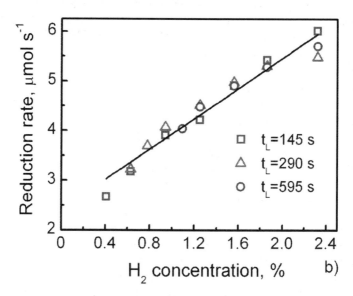

Figure 12. Linear relation between overall NO$_x$ reduction rate *vs.* hydrogen concentraton fed during rich period.

Finally, the term "operation map" is suggested for the set of curves represented in Fig. 11, as a tool for finding any combination of the three studied operational variables: the duration of the lean period, the duration of the rich period and the concentration of the reducing agent: to run the NSR process efficiently. Two ideas may arise from this map. First, every manufactured catalyst can be associated with its own map, so that the comparison of maps will provide information about their relative efficiency when running under the real application. Secondly, one may wonder if any operation point in the map of Fig. 11 is susceptible to be chosen as the best combination (CH2,tR,tL)(CH2,tR,tL)to run in real application.

5.3. Performance Of No_x Storage-Reduction Catalyst In The Temperature-Reductant Concentration Domain By Response Surface Methodology

All previous experiments were carried out at the temperature of 330 ºC, at which N_2O at the reactor exit was negligible, thus being selectivity distributed between N_2 and NH_3, depending on the lean and rich period durations and the hydrogen concentration during the rich period. In this section, the NSR performance trends of the $Pt-Ba/Al_2O_3$ monolith catalyst will be studied at different temperatures and varying the hydrogen concentration fed during the regeneration period by the response surface methodology (RSM). The NO_x storage and reduction behaviour was tested over 9 levels of temperature: 100, 140, 180, 220, 260, 300, 340, 380 and 420 ºC and 9 levels of hydrogen concentration: 0.4, 0.55, 0.7, 0.85, 1, 1.5, 2, 2.5 and 3 % [57].

Fig. 13a shows the NO_x storage capacity (μSTOμSTO) response surface in the hydrogen concentration and reactor inlet temperature domain. With the aim of finding the optimal region, isocurves corresponding to different levels of μSTOμSTO projected to the T-CH2T-CH2space are drawn in Fig. 13b. In the region comprised between temperatures of 220 and 260 ºC and hydrogen concentrations of 1.75 and 3 % a nearly flat surface corresponding to the maximum NO_x storage capacity above 80 % is observed (shaded region). These optimal operational conditions correspond to intermediate temperature and excess of hydrogen. At lower temperatures the conversion of NO to NO_2 was not favoured whereas at higher temperatures the stability of the stored nitrates was reduced leading in both cases to a decrease in the NO_x storage capacity [85]. On the other hand, operating with low hydrogen concentration (<1 % H_2) resulted also in a sharp decrease in the NO_x storage capacity due to the incomplete regeneration of the catalyst [56].

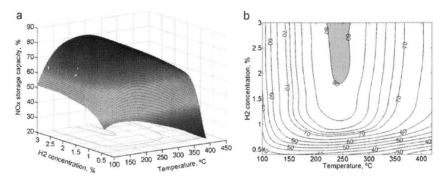

Figure 13. A) NO_x storage capacity response surface in the temperature and hydrogen dose domain, and (b) isocurves corresponding to different levels of trapping efficiencies projected to the T–C_{H2} space.

Likewise, Fig. 14a shows the selectivity to nitrogen response surface in the hydrogen concentration and temperature domain and Fig. 14b the projected iso-selectivity to nitrogen curves. At low temperature (<150 ºC) the selectivity to nitrogen resulted nearly independent of the hydrogen concentration as almost vertical lines can be observed. In this region, the product selectivity changed from N_2O at low H_2 concentrations to NH_3 at higher ones, but remaining practically constant the selectivity to nitrogen. For example, at 100 ºC, $N_2O/NH_3/N_2$ = 54.5/5.1/40.4 for 0.4 % H_2; $N_2O/NH_3/N_2$ = 34.7/33.1/33.2 for 1 % H_2; $N_2O/NH_3/N_2$ = 19.5/48.8/31.7 for 3 % H_2. For higher temperatures (>180 ºC), where the formation of N_2O was negligible, the influence of hydrogen concentration on the selectivity to nitrogen became markedly significant. In this region, the higher hydrogen concentration the lower nitrogen selectivity, and therefore the higher ammonia, was obtained. For example, at 340 ºC, $N_2O/NH_3/N_2$ = 3.4/2.2/94.5 for 0.4 % H_2; $N_2O/NH_3/N_2$ = 0.8/9.4/89.9 for 1 % H_2; $N_2O/NH_3/N_2$ = 0.6/25.0/74.4 for 3 % H_2.

The optimal operational window which resulted in a selectivity to nitrogen higher than 90 % was situated at intermediate-high temperatures (T >250 ºC) and low hydrogen concentrations (CH2CH2 < 1 %) as it can be seen in Fig. 14b (shaded region).

The optimal operating region for maximizing NO_x trapping efficiency and nitrogen selectivity, Fig. 13b and 14b, respectively, did not intercept to each other. The first was maximum at intermediate temperatures and high H_2 concentrations, T = 220–260 ºC and C_{H2} = 1.75–3 % H_2; whereas the selectivity to nitrogen was favoured by high temperatures and low H_2 concentrations, T > 250 ºC and CH2CH2 < 1 %.

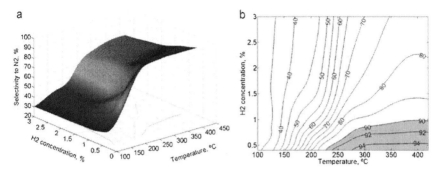

Figure 14. A) Selectivity to N_2 response surface in the temperature and hydrogen dose domain, and (b) isocurves corresponding to different levels of nitrogen selectivity projected to the T–C_{H2} space.

In conventional NSR systems NO_x conversion towards N_2 should be maximized while NH_3 formation should be avoided. The percentage of NO_x converted into nitrogen relative to the total amount of NO entering the trap has been defined as global NSR efficiency (εNSRεNSR, eqn. 23), which allows one to look for the optimal combination of reactor inlet temperature and hydrogen concentration during rich period to obtain the most efficient NSR operation. Fig. 15a shows the εNSRεNSR surface response in the hydrogen concentration and temperature domain and Fig. 15b the projected iso-efficiency curves. As it could be expected, an optimal interval of temperature and hydrogen concentration situated between the optimal operational conditions to obtain maximum storage and maximum selectivity was found. The NSR efficiency resulted higher than 60 %, that is, more than 60 % of NO entering the trap was converted into nitrogen, in the following operating window: T = 250–350 ºC and CH2CH2 = 0.8–1.5. This region provides the best compromise between NO_x storage capacity and selectivity to N_2 to maximize the nitrogen production at the reactor exit related to the total amount of NO at the reactor inlet. The most efficient operation corresponds to 0.9 % H_2 and 280 ºC reaching 65 % of global NSR efficiency (Fig. 15b).

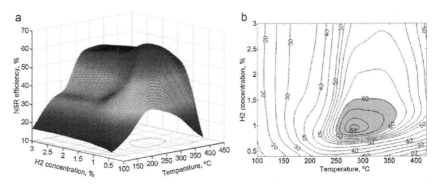

Figure 15. A) Global NSR efficiency response surface in the temperature and hydrogen dose domain, and (b) isocurves corresponding to different levels of NSR efficiencies projected to the T–C_{H2} space.

6. CONCLUSIONS

The NO_x storage and reduction technology for diesel eshaust aftertreatment has been studied in the present chapter, including the synthesis of Pt-BaO/Al_2O_3 monolith catalyst, the involved reaction mechanisms or chemistry of the process, and the control of engineering parameters, of importance in the real application in automobiles, to remove nitrogen oxides most efficiently by conversion to nitrogen.

The preparation methodology for Pt-BaO/Al_2O_3 monolith catalyst has been described. The monolith is primarily washcoated with a thin film of porous alumina. Then platinum is incorporated by adsorption (ion exchange) from a Pt(NH_3)$_4$(NO_3)$_2$ aqueous solution. Finally, the barium as NO_x storage component is incorporated by dry impregnation from a Ba(CH_3-COO)$_2$ aqueous solution. This procedure achieves homogeneous distribution as well as high dispersion of platinum and barium on the catalyst surface, then providing the adequate Pt-Ba proximity which enhances the interaction needed for Pt to promote both the initial NO oxidation and the reduction of N_xO_y adspecies on the Ba sites.

The chemistry of NO_x regeneration and reduction mechanisms has been reviewed. Operando FTIR experiments of NO_x adsorption on powder Pt-Ba/Al_2O_3 samples has shown that below 250 ºC nitrite species are predominant whereas above 250 ºC nitrate species are predominant. Thus, two parallel routes have been verified. The "nitrite route" where NO is oxidized on Pt sites and stored onto Ba neighbouring sites in the form of nitrite ad-species which can progressively transform into nitrates depending on the reaction temperature. The second route, called "nitrate route" implies the oxidation of NO to NO_2 on Pt, then NO_2 desproportionation on Ba to form nitrates NO evolved into the gas phase.

During the regeneration period, when oxygen is shifted by hydrogen, the reduction of stored nitrites and nitrates leads to the formation of different nitrogen containing species, namely N_2, N_2O and NH_3 along with water. Nitrogen formation involves first the fast formation of ammonia by reaction of nitrates with H_2 and then the subsequent conversion of the ammonia formed with stored nitrates leading to the selective formation of N_2. At temperature above 330 ºC, N_2O was almost negligible.

In the automobile practice, the operational conditions at which the process is conducted affect significantly the NSR behaviour of a Pt-BaO/Al_2O_3 monolith catalyst, such as the duration of lean and rich periods and the concentration of reductant fed during the regeneration period. There exists a given amount of hydrogen which is needed to achieve the complete reduction of NO_x (stored during the lean period and fed during the rich period). Below that minimum, as regeneration is not complete the reduction conversion is lower and consequently the storage capacity in the subsequent lean period is also reduced. However, with hydrogen in defect very high selectivity towards N_2 is achieved. On the other hand, with hydrogen in excess the formation of ammonia increases notably, although NO_x storage capacity is practically maintained at maximum and almost total reduction conversion is achieved. The maximum global NSR efficiency (percentage of N_2 at the exit related to NO at the entry) is achieved just at the stoichiometric point, when the amount of H_2 is neither in defect nor in excess. The amount of hydrogen fed during the rich period is proportional to the product CH2×tRCH2×tRso that this amount can be controlled by managing either H_2 concentration or duration of the regeneration period. In fact, there exist different combinations (CH2,tR)(CH2,tR) which achieve similar NSR efficiency. The locus of these combinations conforms the isoeffiency curve map with NSR efficiency as the response parameter.

The combined analysis of temperature (100-420 ºC) and H_2 concentration (0.4-3 %), maintaining lean and rich period times in 145 and 25 s respectively, has allowed to find the maximum storage capacity at intermediate temperature (~240 ºC) and high reductant concentration (>2 % H_2). Maximum selectivity to N_2 has been obtained operating at high temperature (>300 ºC) and hydrogen in defect (<1 % H_2). The optimal control is performed at intermediate position, i.e. 270 ºC and 1 % H_2, at which the maximum global NSR efficieny is achieved.

NOMENCLATURE

ABREVIATIONS

ADS	Adsorption from solution, catalyst preparation procedure.
DI	Dry impregnation
FTIR	Fourier Transform Infrared spectroscopy.
ICP-MS	Inductively Coupled Plasma Mass Spectroscopy.
LNT	Lean NO_x Trap.
MS	Mass Spectroscopy.
NSR	NO_x Storage and Reduction.
PM	Particulate matter.
RSM	Response Surface Methodology.
SEM-EDX	Scanning Electronic Microscopy-Energy Dispersed X-Ray Spectroscopy.
SCR	Selective Catalytic Reduction.
STP	Standard temperature and pressure
TEM	Transmission Electronic Microscopy.
TWC	Three way catalyst.
WI	Wet impregnation, catalyst preparation procedure.

ACKNOWLEDGEMENT

The authors wish to acknowledge the financial support provided by the Spanish Science and Innovation Ministry (CTQ2009-125117) and the Basque Government (Consolidated Research Group, GIC 07/67-JT-450-07).

REFERENCES

1. B Pereda-ayo, NOx storage and reduction (NSR) for diesel engines: synthesis of Pt-Ba/Al2O3 monolith catalyst, reaction mechanisms and optimal control of the process. PhD Thesis, University of the Basque Country, Bilbao, Spain; 2012
2. G Mazzarella, F Ferraraccio, M. V Prati, S Annunziata, A Bianco, A Mezzogiorno, G Liguori, I. F Angelillo, M Cazzola, Effects of diesel exhaust particles on human lung epithelial cells: An in vitro study. Respiratory Medicine 20071011155
3. R. G Derwent, The long-range transport of ozone within Europe and its control. Environmental Pollution 1990299318
4. V. I Parvulescu, P Grange, B Delmon, Catalytic removal of NO. Catalysis Today 199846233
5. L. J Folinsbee, Human health-effects of air-pollution. Environmental Health Perspectives 199310045
6. S. V Krupa, R. N Kickert, The greenhouse effect- impacts of ultraviolet-B (Uv-B) radiation, carbon-dioxide (CO2), and ozone (O3) on vegetation. Environmental Pollution 198961263
7. F Klingstedt, K Arve, K Eranen, D. Y Murzin, Toward improved catalytic low-temperature NOx removal in diesel-powered vehicles. Accounts of Chemical Research 200639273
8. DieselnetEmission standards. European Union. Cars and light trucks. http://www.dieselnet.com/standards/eu/ld.phpaccessed 1 May 2012
9. K. C Taylor, Nitric-oxide catalysis in automotive exhaust systems. Catalysis Reviews Science Eng. 199335457
10. K Tamaru, G. A Mills, Catalysts for control of exhaust emissions. Catalysis Today 199422349
11. R. M Heck, R. J Farrauto, Automobile exhaust catalysts. Applied Catalysis A: General 2001221443
12. J. R González-velasco, J Entrena, J. A González-marcos, J. I Gutiérrez-ortiz, M. A Gutiérrez-ortiz, Preparation, activity and durability of promoted platinum catalysts for automotive exhaust control. Applied Catalysis B: Environmental 19943191
13. J. G Numan, H. J Robota, M. J Cohn, S. A Bradley, Physicochemical properties of Ce-containing three-way catalysts and the effect of Ce on catalyst activity. Journal of Catalysis 1992133309

14. B. H Engler, D Lindner, E. S Lox, Schäfer-Sindlinger, Ostgathe K. Development of improved Pd-only and Pd/Rh three-way catalysts. Studies in Surface Science and Catalysis 199596441

15. R. M Heck, R. J Farrauto, S Gulati, Catalytic Air Pollution Control: Commercial Technology. New Yersey: John Wiley & Sons; 2009

16. Seijger GBFCerium-ferrierite catalyst systems for reduction of NOx in lean burn engine exhaust gas. PhD Thesis, Technical University Delft, Delft, Netherlands; 2002

17. J. R González-velasco, M. P González-marcos, M. A Gutiérrez-ortiz, J. A Botas-echevarría, Catálisis, automóvil y medio ambiente. Anales de Química 2002424

18. F Basile, G Fomasari, A Grimandi, M Livi, A Vaccari, Effect of Mg, Ca and Ba on the Pt-catalyst for NOx storage reduction. Applied Catalysis B: Environmental 20066958

19. J. M Trichard, Current tasks and challenges for exhaust aftertreatment research: An industrial viewpoint. Studies in Surface Science and Catalysis 2007171211

20. G Centi, S Perathoner, Introduction: State of the art in the development of catalytic processes for the selective catalytic reduction of NOx into N2. Studies in Surface Science and Catalysis 20071711

21. T. V Johnson, Review of diesel emissions and control. International Journal of Engine Research 200910275

22. M. M Maricq, Chemical characterization of particulate emissions from diesel engines: A review. Journal of Aerosol Science 2007381079

23. D Fino, Diesel emission control: Catalytic filters for particulate removal. Science and Technology of Advanced Materials 2007893

24. S Biswas, V Verma, J. J Schauer, C Sioutas, Chemical speciation of PM emissions from heavy-duty diesel vehicles equipped with diesel particulate filter (DPF) and selective catalytic reduction (SCR) retrofits. Atmospheric Environment 2009431917

25. S Brandenberger, O Krocher, A Tissler, R Althoff, The state of the art in selective catalytic reduction of NOx by ammonia using metal-exchanged zeolite catalysts. Catalysis Reviews, Science and Engineering 200850492

26. M. D Amiridis, T. J Zhang, R. J Farrauto, Selective catalytic reduction of nitric oxide by hydrocarbons. Applied Catalysis B: Environmental 199610203
27. N Takahashi, H Shinjoh, T Iijima, T Suzuki, K Yamazaki, K Yokota, H Suzuki, N Miyoshi, S Matsumoto, T Tanizawa, T Tanaka, S Tateishi, K Kasahara, The new concept 3way catalyst for automotive lean-burn engine: NOx storage and reduction catalyst. Catalysis Today 1996
28. S. I Matsumoto, Recent advances in automobile exhaust catalysts. Catalysis Today 200490183
29. D. N Belton, K. C Taylor, Automobile exhaust emission control by catalysts. Current Opinion in Solid State & Materials Science 1999497
30. U. G Alkemade, B Schumann, Engines and exhaust after treatment systems for future automotive applications. Solid State Ionics 20061772291
31. M. V Twigg, Progress and future challenges in controlling automotive exhaust gas emissions. Applied Catalysis B: Environmental 2007702
32. E Fridell, M Skoglundh, S Johansson, B. R Westerberg, A Törncrona, G Smedler, Investigations of NOx storage catalysts. Studies in Surface Science and Catalysis 1998116537
33. N Miyoshi, S Matsumoto, T Katoh, T Tanaka, J Harada, N Takahashi, K Yokota, M Sugiara, K Kasahara, Development of new concept three-way catalyst for automotive lean-burn engines. SAE Technical Paper Series 950809; 1995
34. W. S Epling, L. E Campbell, A Yezerets, N. W Currier, J. E Parks, Overview of the fundamental reactions and degradation mechanisms of NOx storage/reduction catalysts. Catalysis Reviews, Science and Engineering 200446163
35. S Roy, A Baiker, NOx Storage-reduction catalysis: from mechanism and materials properties to storage-reduction performance. Chemical Reviews 20091094054
36. G Liu, P. X Gao, A review on NOx storage/reduction catalysts: mechanism, materials and degradation studies. Catalysis Science & Technology 20111552
37. Y. J Li, S Roth, J Dettling, T Beutel, Effects of lean/rich timing and nature of reductant on the performance of a NOx trap catalyst. Topics in Catalysis 200116139

38. W. S Epling, A Yezerets, N. W Currier, The effect of exothermic reactions during regeneration on the NOx trapping efficiency of a NOx storage/reduction catalyst. Catalysis Letters 2006110143

39. H Abdulhamid, E Fridell, M Skoglundh, The reduction phase in NOx storage catalysis: Effect of type of precious metal and reducing agent. Applied Catalysis B: Environmental 200662319

40. V Meille, Review on methods to deposit catalysts on structured surfaces. Applied Catalysis A: General 20063151

41. T. A Nijhuis, Beers AEW, Vergunst T, Hoek I, Kapteijn F, Moulijn JA. Preparation of monolithic catalysts. Catalysis Reviews, Science and Engineering 200143345

42. P Avila, M Montes, E Miro, Monolithic reactors for environmental applications- A review on preparation technologies. Chemical Engineering Journal 200510911

43. M Piacentini, M Maciejewski, A Baiker, NOx storage-reduction behavior of Pt-Ba/MO2 (MO2 = SiO2, CeO2, ZrO2) catalysts. Applied Catalysis B: Environmental 200772105

44. I Malpartida, Vargas MAL, Alemany LJ, Finocchio E, Busca G. Pt-Ba-Al2O3 for NOx storage and reduction: Characterization of the dispersed species. Applied Catalysis B: Environmental 200880214

45. I Nova, L Lietti, P Forzatti, Mechanistic aspects of the reduction of stored NOx over Pt-Ba/Al2O3 lean NOx trap systems. Catalysis Today 2008136128

46. C Agrafiotis, A Tsetsekou, The effect of powder characteristics on washcoat duality. Part I: Alumina washcoats. Journal of the European Ceramic Society 200020815

47. B Pereda-ayo, R López-fonseca, J. R González-velasco, Influence of the preparation procedure of NSR monolithic catalysts on the Pt-Ba dispersion and distribution. Applied Catalysis A: General 200936373

48. A Tsetsekou, C Agrafiotis, A Milias, Optimization of the rheological properties of alumina slurries for ceramic processing applications- Part I: Slip-casting. Journal of the European Ceramic Society 200121363

49. M Valentini, G Groppi, C Cristiani, M Levi, E Tronconi, P Forzatti, The deposition of gamma-Al2O3 layers on ceramic and metallic supports for the preparation of structured catalysts. Catalysis Today 200169307

50. C Agrafiotis, A Tsetsekou, Deposition of meso-porous gamma-alumina coatings on ceramic honeycombs by sol-gel methods. Journal of the European Ceramic Society 200222423

51. A Lindholm, N. W Currier, J Dawody, A Hidayat, J Li, A Yezerets, L Olsson, The influence of the preparation procedure on the storage and regeneration behavior of Pt and Ba based NOx storage and reduction catalysts. Applied Catalysis B: Environmental 200988240

52. B Pereda-ayo, D Duraiswami, R López-fonseca, J. R González-velasco, Influence of platinum and barium precursors on the NSR behavior of Pt-Ba/Al2O3 monoliths for lean-burn engines. Catalysis Today 2009147244

53. J. R Regalbuto, K Agashe, A Navada, M. L Bricker, Q Chen, A scientific description of Pt adsorption onto alumina. Studies in Surface Science and Catalysis 1998118147

54. W. A Spieker, J. R Regalbuto, A fundamental model of platinum impregnation onto alumina. Chemical Engineering Science 2001563491

55. J. R Regalbuto, A Navada, S Shadid, M. L Bricker, Q Chen, An experimental verification of the physical nature of Pt adsorption onto alumina. Journal of Catalysis 1999184335

56. B Pereda-ayo, D Duraiswami, J. J Delgado, R López-fonseca, J. J Calvino, S Bernal, J. R González-velasco, Tuning operational conditions for efficient NOx storage and reduction over a Pt-Ba/Al2O3 monolith catalyst. Applied Catalysis B: Environmental 201096329

57. B Pereda-ayo, D Duraiswami, J. A González-marcos, J. R González-velasco, Performance of NOx storage-reduction catalyst in the temperature-reductant concentration domain by response surface methodology. Chemical Engineering Journal 201116958

58. N. W Cant, Liu IOY, Patterson MJ. The effect of proximity between Pt and BaO on uptake, release, and reduction of NOx on storage catalysts. Journal of Catalysis 2006243309

59. R. D Clayton, M. P Harold, V Balakotaiah, C. Z Wan, Pt dispersion effects during NOx storage and reduction on Pt/BaO/Al2O3 catalysts. Applied Catalysis B: Environmental 200990662

60. I Nova, L Castoldi, L Lietti, E Tronconi, P Forzatti, F Prinetto, G Ghiotti, NOx adsorption study over Pt-Ba/alumina catalysts: FT-IR and pulse experiments. Journal of Catalysis 2004222377

61. I Nova, L Castoldi, L Lietti, E Tronconi, P Forzatti, On the dynamic behavior of "NOx-storage/reduction" Pt-Ba/Al2O3 catalyst. Catalysis Today 200275431

62. L Lietti, P Forzatti, I Nova, E Tronconi, NOx Storage Reduction over Pt---Ba/[gamma]-Al2O3 Catalyst. Journal of Catalysis 2001204175

63. F Prinetto, G Ghiotti, I Nova, L Lietti, E Tronconi, P Forzatti, FT-IR and TPD investigation of the NOx storage properties of BaO/Al2O3 and Pt-BaO/Al2O3 catalysts. Journal of Physical Chemistry B 200110512732

64. I Nova, L Castoldi, F Prinetto, Dal Santo V, Lietti L, Tronconi E, Forzatti P, Ghiotti G, Psaro R, Recchia S. NOx adsorption study over Pt-Ba/alumina catalysts: FT-IR and reactivity study. Topics in Catalysis 2004

65. P Forzatti, L Castoldi, I Nova, L Lietti, E Tronconi, NOx removal catalysis under lean conditions. Catalysis Today 2006117316

66. E Fridell, M Skoglundh, B Westerberg, S Johansson, G Smedler, NOx storage in barium-containing catalysts. Journal of Catalysis 1999183196

67. P Broqvist, H Gronbeck, E Fridell, I Panas, NOx storage on BaO: theory and experiment. Catalysis Today 20049671

68. B. R Westerberg, E Fridell, A transient FTIR study of species formed during NOx storage in the Pt/BaO/Al2O3 system. Journal of Molecular Catalysis A: Chemical 2001165249

69. J Szanyi, J. H Kwak, J Hanson, C. M Wang, T Szailer, Peden CHF. Changing morphology of BaO/Al2O3 during 2uptake and release. Journal of Physical Chemistry B 20051097339

70. U Elizundia, R Lopez-fonseca, I Landa, M. A Gutierrez-ortiz, J. R Gonzalez-velasco, FT-IR study of NOx storage mechanism over Pt/BaO/Al2O3 catalysts. Effect of the Pt-BaO interaction. Topics in Catalysis 2007

71. H Mahzoul, J. F Brilhac, P Gilot, Experimental and mechanistic study of NOx adsorption over NOx trap catalysts. Applied Catalysis B: Environmental 19992047

72. N. W Cant, M. J Patterson, The storage of nitrogen oxides on alumina-supported barium oxide. Catalysis Today 200273271

73. S Kikuyama, I Matsukuma, R Kikuchi, K Sasaki, K Eguchi, A role of components in Pt-ZrO2/Al2O3 as a sorbent for removal of NO and 2Applied Catalysis A: General 200222623

74. F Rodrigues, L Juste, C Potvin, J. F Tempere, G Blanchard, G Djega-mariadassou, NOx storage on barium-containing three-way catalyst in the presence of CO2. Catalysis Letters 20017259

75. L Olsson, H Persson, E Fridell, M Skoglundh, B Andersson, Kinetic study of NO oxidation and NOx storage on Pt/Al2O3 and Pt/BaO/Al2O3. Journal of Physical Chemistry B 20011056895

76. L Olsson, B Westerberg, H Persson, E Fridell, M Skoglundh, B Andersson, A kinetic study of oxygen adsorption/desorption and NO oxidation over Pt/Al2O3 catalysts. Journal of Physical Chemistry B 199910310433

77. X. G Li, M Meng, P. Y Lin, Y. L Fu, T. D Hu, Y. N Xie, J Zhang, Study on the properties and mechanisms for NOx storage over Pt/BaAl2O4Al2O3 catalyst. Topics in Catalysis 2003

78. L Cumaranatunge, S. S Mulla, A Yezerets, N. W Currier, W. N Delgass, F. H Ribeiro, Ammonia is a hydrogen carrier in the regeneration of Pt/BaO/Al2O3 NOx traps with H2. Journal of Catalysis 200724629

79. V Medhekar, V Balakotaiah, M. P Harold, TAP study of NOx storage and reduction on Pt/Al2O3 and Pt/Ba/Al2O3. Catalysis Today 2007121226

80. Z. Q Liu, J. A Anderson, Influence of reductant on the thermal stability of stored NOx in Pt/Ba/Al2O3 NOx storage and reduction traps. Journal of Catalysis 200422418

81. S. S Mulla, S. S Chaugule, A Yezerets, N. W Currier, W. N Delgass, F. H Ribeiro, Regeneration mechanism of Pt/BaO/Al2O3 lean NOx trap catalyst with H2", Catalysis Today 2008136136

82. W. P Partridge, J. S Choi, N. H Formation, and utilization in regeneration of Pt/Ba/Al2O3 NOx storage-reduction catalyst with H2. Applied Catalysis B: Environmental 200991144

83. I Nova, L Lietti, L Castoldi, E Tronconi, P Forzatti, New insights in the NOx reduction mechanism with H2 over Pt-Ba/gamma-Al2O3 lean NOx trap catalysts under near-isothermal conditions. Journal of Catalysis 2006239244

84. B Pereda-ayo, J. R González-velasco, R Burch, C Hardacre, S Chansai, Regeneration mechanism of a Lean NOx Trap (LNT) catalyst in the presence of NO investigated using isotope labelling techniques. Journal of Catalysis 2012285177

85. L Lietti, I Nova, P Forzatti, Role of ammonia in the reduction by hydrogen of NOx stored over Pt-Ba/Al2O3 lean NOx trap catalysts. Journal of Catalysis 2008257270

86. R. D Clayton, M. P Harold, V Balakotaiah, NOx storage and reduction with H2 on Pt/BaO/Al2O3 monolith: Spatio-temporal resolution of product distribution. Applied Catalysis B: Environmental 200884616

87. A Lindholm, N. W Currier, E Fridell, A Yezerets, L Olsson, NOx storage and reduction over Pt based catalysts with hydrogen as the reducing agent. Influence of H2O and CO2. Applied Catalysis B: Environmental 20077578

88. W. S Epling, A Yezerets, N. W Currier, The effects of regeneration conditions on NOx and NH3 release from NOx storage/reduction catalysts. Applied Catalysis B: Environmental 200774117

89. I Nova, L Castoldi, L Lietti, E Tronconi, P Forzatti, How to control the selectivity in the reduction of NOx with H2 over Pt-Ba/Al2O3 Lean NOx Trap catalysts. Topics in Catalysis 2007

90. K. S Kabin, R. L Muncrief, M. P Harold, Y. J Li, Dynamics of storage and reaction in a monolith reactor: lean NOx reduction. Chemical Engineering Science 2004595319

91. R. D Clayton, M. P Harold, V Balakotaiah, Performance Features of Pt/BaO Lean NOx Trap with Hydrogen as Reductant. Aiche Journal 200955687

Analysis of Two Stroke Marine Diesel Engine Operation Including Turbocharger Cut-Out by Using a Zero-Dimensional Model

Cong Guan [1], Gerasimos Theotokatos [2,*] and Hui Chen[1]

[1]Key Laboratory of High Performance Ship Technology of Ministry of Education, School of Energy and Power Engineering, Wuhan University of Technology, 1178 Heping Road, Wuhan 430063, China

[2]Department of Naval Architecture, Ocean & Marine Engineering, University of Strathclyde, 100 Montrose Street, Glasgow G4 0LZ, UK

ABSTRACT

In this article, the operation of a large two-stroke marine diesel engine including various cases with turbocharger cut-out was thoroughly investigated by using a modular zero-dimensional engine model built in MATLAB/Simulink environment. The model was developed by using as a basis an in-house modular mean value engine model, in which the existing cylinder block was replaced by a more detailed one that is capable of representing the scavenging ports-cylinder-exhaust valve processes. Simulation of the engine operation at steady state conditions was performed and the derived engine performance parameters were compared with the respective values obtained by the engine shop trials. The investigation of engine operation under turbocharger cut-out conditions in the region from 10% to 50% load was carried out and the influence of turbocharger cut-out on engine performance including the in-cylinder parameters was comprehensively studied. The recommended schedule for the combination of the turbocharger cut-out and blower activation was discussed for the engine operation under part load conditions. Finally, the influence of engine operating strategies on the annual fuel savings, CO_2 emissions reduction and blower operating hours for a Panamax container ship operating at slow steaming conditions is presented and discussed.

Keywords: two-stroke marine diesel engine; zero-dimensional model; part load operation; turbocharger cut-out; blower activation

1. INTRODUCTION

During the last years the maritime industry has confronted multiple issues stemming from different factors, such as the increased bunker prices [1,2,3], the significant reduction of chartered ship rates [4], as well as the oversupply of shipping transport capacity [5,6,7], which have put huge financial pressure to the shipping companies. Furthermore, the growing concern for suppressing emissions from shipping along with the recent more stringent international and national regulations for limiting greenhouse and non-greenhouse emissions [8,9,10] have set additional challenges. Therefore, ship owners and operators are forced to adopt measures to lower fuel consumption and the associated costs as well as to reduce the ship gaseous emissions.

In order to achieve a more efficient and environmentally friendly ship operation, a number of measures have been proposed and used. These include the introduction of the electronically controlled versions of marine diesel engines [11,12], in which the engine settings (fuel injection and exhaust valve opening/closing timings) can be controlled and thus the engine can operate in various modes with high efficiency and low emission throughout the entire operating envelope, the application of the exhaust gas bypass, the usage of turbochargers with variable geometry turbines [13,14] and the installation of waste heat recovery systems employing steam turbine to generate electricity (in some cases combined with a power turbine) [15,16,17]. For the existing ships, retrofitting engine packages for fuel slide valves and cylinder lubricators can be used; the former to improve engine injection and combustion processes; the latter to optimise the cylinder oil consumption avoiding the negative effects of overdosing of alkaline cylinder lubrication [18,19]. Apart from the above mentioned engine related measures, the improvement of vessel hydrodynamic performance (optimum hull form, low resistance coating and effective hull and propeller interaction, *etc.*) [20] as well as the introduction of propeller saving devices [21] can result in potential fuel savings.

However, the majority of the technical measures mentioned above require high investment cost achieving limited fuel consumption reduction potential. Therefore, operational measures including hull/propeller maintenance management, voyage planning, trim optimisation and operation of engine at slow steaming conditions have also been proposed. Slow steaming has been adopted in shipping industry as a standard operating strategy in order for large containerships to stay profitable in the competitive market, since it is the most immediate and effective way to lower operational cost associated with fuel consumption as well as emissions [22,23]. A significant number of ship operators have even implemented ultra-slow steaming [24] in order to maximize the expected benefits. Even considering the last year reduction of fuel prices, the vessels' operational speed is not expected to increase, since the fuel cost is an important contributor to the shipping company profitability and in addition, containerships with reduced design speed have been recently built, whilst retrofits of bulbous bow have been applied in many containerships to render the ship hull more efficient at lower speeds.

In spite of the substantial reduction of fuel consumption, slow steaming results in less efficient engine operation and introduces some challenges for the engine systems, as the engine operating at low loads results in low exhaust gas energy content supplied to the turbocharger turbine, lower turbocharger speed and compressor air flow. This can lead to increased deposits at the cylinder and exhaust system components. The low temperatures at combustion result in low wall temperatures especially in the cylinder liner bottom part, which can induce cold corrosion problems. Therefore, attention is needed so that slow steaming does not limit the fuel saving potential and cause engine damage considering that deposits accumulate after a long period of engine slow steaming operation [25]. In this respect, solutions including the increase of engine jacket cooling water temperature and the temporal engine operation at higher load have been proposed by the engine manufacturers for ensuring an efficient and reliable engine operation down to 10% load [26]. For the two stroke diesel engines with two or more turbochargers, turbochargers cut-out, which has already been used by many ship owners, is a very viable option for further optimising fuel consumption and improving engine part load operation [27]. However, the turbochargers cut-out and especially, the cut-out of one unit out of two or two units out of three may substantially increase the engine cylinders maximum pressure when the engine operates at the low load region, and as a consequence, may cause engine structural issues. Therefore, the systematic investigation of the engine operation in such cases is quite crucial for understanding the interactions of the various engine components as well as for obtaining the engine performance.

1. Introduction

Due to the marine diesel engine high cost and large size, appropriate simulation tools with varying degree of complexity have been used for investigating the engine steady-state performance and transient response as well as for assisting the design process of engine and its systems. The most commonly used types are the cycle mean value engine models (MVEM) [28,29,30,31,32] and the zero-dimensional models [33,34,35,36,37]. The former are fast running and need less input, but they require an elaborate setting up phase in order to predict the engine behaviour with sufficient accuracy. Furthermore, the in-cycle variation (per degree crank angle) cannot be represented [38,39] and therefore the modelling of engines with varying settings poses difficulties since parametric response surfaces are needed [40,41]. The latter are more complex and require a greater amount of input data and execution time, but they can represent the engine working processes more accurately and predict the in-cylinder parameters variation.

Although slow steaming is commonly used in the last years, only limited published works have investigated this specific engine operating phases [41,42,43]. Kyrtatos *et al.* [43] proposed a novel way of evaluating engine performance by comparing service monitored data and thermodynamic model predictions, and they carried out a study for predicting a VLCC main engine performance at slow steaming conditions concluding that a sufficient extrapolation of the compressor map is required in order to avoid errors in simulation results. Hountalas *et al.* [42] studied the effect of one turbocharger unit cut-out (out of two) on the engine performance and it is concluded that retard of fuel injection could be the solution for reducing the cylinder maximum pressure and the associated NOx emissions. In [41], a MVEM was used to investigate the engine performance at low load operation considering the influence of blower activation/deactivation phases and turbocharger cut-out. A reliable method of extrapolating the compressor map at low speed area was developed and validated. However, it was found that the BSFC variation cannot be captured using MVEM for engine with varying settings including turbocharger cut-out, as the MVEM parameters were calibrated only for normal operation and the in-cylinder parameters could not be assessed due to the limitation of MVEM approach.

This work focuses on the thorough investigation of a two stroke marine diesel engine with emphasis at part load operating conditions using a zero-dimensional model, which was developed according to a modular approach in MATLAB/Simulink computational environment. As an effective technique to improve engine performance at low loads, turbocharger cut-out is studied in the region from 10% to 50% load. Various options are investigated including cut-out of one or two turbocharger units (out of three) in combination with blower activation/deactivation. The calculated engine performance parameters as well as the in-cylinder parameters variation are analysed and the engine behaviour is explained. Based on the presented results, recommendations are discussed for the required number of turbochargers cut-out in conjunction with blower activation/deactivation when the investigated engine operates at slow steaming conditions within the load range from 10% to 50% of the MCR point. In addition, the annual figures in terms of fuel consumption, CO_2 emissions and blower operating hours for a panamax size containership are estimated for various engine operating modes. Based on the derived results, the benefits and savings of the proposed operating mode are identified and discussed.

2. MODULAR ENGINE MODEL DESCRIPTION

2.1. Model Structure

In this work, a zero-dimensional model implemented in MATLAB/Simulink environment as shown in Figure 1a is used for investigating a large two-stroke diesel engine operation. The model was developed based on the structure of a modular mean value engine model presented previously [39,41]. The difference from the respective MVEM is that the cylinder block shown in Figure 1a includes the blocks for modelling the cylinder processes, the scavenging ports and exhaust valves. The cylinder block was developed based on a zero-dimensional approach. The part of cylinder block representing the cylinder No. 1 including the scavenging ports and the exhaust valve is shown in Figure 1b.

According to the modular modelling structure, the engine components are modelled by using individual blocks capable of representing the involved processes. Connections between the adjacent blocks are used for exchanging the required parameters. Flow receiver elements (or as otherwise called control volumes according to [44,45]) are used for modelling the engine cylinders, scavenging receiver and exhaust gas receiver. The scavenging ports, exhaust valves as well as the turbocharger compressor and turbine are considered as flow elements. For modelling the engine boundaries, the concept of fixed parameter blocks is used, in which the working medium state as described by the temperature, pressure and equivalence ratio remains unchanged. Shaft elements are employed to represent the engine crankshaft and turbocharger shafts. The proportional-integral (PI) controller block combined with sub-blocks representing the engine fuel rack limiters is used to model the engine governor. The propeller block is used for representing the ship propeller behaviour. The working medium is assumed to be homogeneous and its properties are calculated considering the corresponding temperature, equivalence ratio and pressure.

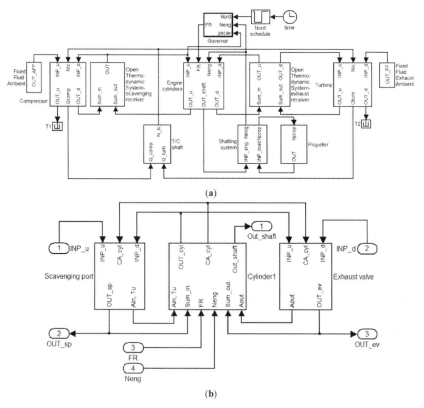

Figure 1. (a) Zero-dimensional two-stroke engine model implemented in MATLAB/Simulink environment; (b) Cylinder No. 1 block along with each scavenging ports and exhaust valves blocks.

The input required for the flow element blocks is taken from the adjacent flow receiver or constant parameter blocks and include the parameters characterising the working medium state (temperature, pressure, composition) and the thermodynamic properties. The mass and energy flows through the flow elements are calculated and provided to the adjacent flow receiver blocks. The absorbed compressor torque and produced turbine torque are additionally calculated in the compressor and turbine elements, respectively, and they subsequently fed to the turbocharger shaft element, which calculates the turbocharger speed. This, in turn, is provided to the turbine and compressor blocks. The crankshaft block calculates the engine and propeller rotational speeds. The engine rotational speed is forwarded to the engine cylinder and governor blocks. The propeller block uses as input the propeller speed and calculates the propeller absorbed torque. The engine crank angle is calculated by integrating the rotational speed and provided as input to engine cylinder block. For solving the used differential equations, an Euler integration scheme was used along with a fixed time step corresponding to approximately 0.5° crank angle.

2.2. Governing Equations

2.2.1. Cylinder

The engine cylinders are modelled using the open or closed thermodynamic system consideration depending on their operating phase. One zone model is considered for the closed cycle and the exhaust gas blowdown period (from EVC to SPO), whereas a two zone model is employed for modelling the scavenging process (from SPO to EVC). Each zone is considered to be uniform and the working medium (air or exhaust gas) is regarded as ideal gas that characterized by its pressure, temperature and equivalence ratio. For the calculation of the cylinder working fluid thermodynamic parameters, the mass and energy conservation laws as well as the ideal gas state equation in the control volume that encloses the cylinder are used [44,45,46]. The one zone model employs three differential equations for calculating the temperature, the mass and burnt fuel fraction along with the ideal gas equation for calculating the pressure and algebraic equations for estimating the working fluid properties.

By applying the energy conservation and the ideal gas state equation assuming that the system can be characterized by using the temperature, pressure and equivalence ratio, the following equation is derived for calculating the cylinder gas temperature time derivative [44]:

$$\frac{dT}{dt} = \frac{B - \frac{\partial u}{\partial \varphi}\frac{d\varphi}{dt} - \frac{p}{D}\frac{\partial u}{\partial p}\left(\frac{1}{R}\frac{\partial R}{\partial \varphi}\frac{d\varphi}{dt} + \frac{1}{m}\frac{dm}{dt} - \frac{1}{V}\frac{dV}{dt}\right)}{\frac{\partial u}{\partial T} + \frac{C\,p}{D\,T}\frac{\partial u}{\partial p}}$$

(1)

where:

$$B = \frac{1}{m}\left(\sum_{sf} \dot{Q}_{sf} + \sum_j \dot{m}_j h_j - u\frac{dm}{dt}\right) - \frac{RT}{V}\frac{dV}{dt}$$

$$C = 1 + \frac{T}{R}\frac{\partial R}{\partial T} \quad \text{and} \quad D = 1 - \frac{p}{R}\frac{\partial R}{\partial p}$$

The above equation is able to consider the dissociation effects occurring at high pressures and temperatures during the combustion phase. The mass time derivative used in Equation (1) is calculated by applying the mass balance equation, whereas the engine cylinders volume derivative is derived by using the engine kinematic mechanism particulars and the cylinder clearance volume [44]. The working medium properties are calculated as functions of temperature, equivalence ratio and pressure as proposed in [45,47].

The burnt fuel fraction (ξ) is defined by the ratio of burnt fuel mass to the total mass. By differentiating the burnt fuel fraction definition equation, the following equation is derived for the burnt fuel fraction time derivative:

$$\frac{d\xi}{dt} = \frac{(\dot{m}\xi)_{in} - (\dot{m}\xi)_{out} + \frac{dm_{fb}}{dt} - \xi\frac{dm}{dt}}{m} \qquad (2)$$

The fuel/air equivalence ratio and its time derivative, used in Equation (1), are calculated by using the burnt fuel fraction and its time derivative, according to the following equations [44]:

$$\varphi = \frac{\xi}{FAs(1-\xi)} \qquad (3)$$

$$\frac{d\varphi}{dt} = \frac{1}{FAs(1-\xi)^2}\frac{d\xi}{dt} \qquad (4)$$

For calculating the fuel burning rate used in Equations (1) and (2), the Woschni-Anisits combustion model [46] was used for describing the combustion process. This model employs the single Vibe approach with the model constants that depend on the engine operating conditions and it can provide adequate accuracy for single fuel combustion [46].

According to that, the fuel burning rate is calculated using the following equation:

$$\frac{dm_{fb}}{dt} = 6N\, m_{f,cy}\, \alpha \frac{(m+1)}{\Delta\theta}\left(\frac{\theta-\theta_{SOC}}{\Delta\theta}\right)^m e^{\left[-\alpha\left(\frac{\theta-\theta_{SOC}}{\Delta\theta}\right)^{m+1}\right]} \quad (5)$$

where $m_{f,cy}$ is the mass of fuel injected per cylinder per cycle, $\Delta\theta$ is the combustion duration, θ_{SOC} is the start of combustion and a, m are the combustion model parameters.

The start of combustion can be calculated by using the start of injection and the ignition delay as follows:

$$\theta_{SOC} = \theta_{SOI} + \Delta\theta_{IGD} \quad (6)$$

The ignition delay for Diesel engine is estimated by using the following Sitkey equation [46]:

$$\Delta\theta_{IGD} = 6N10^{-3}\left[k_1 + k_2 e^{\frac{7800}{69167 RT}}\left(1.0197 p^{-0.7}\right) + k_3 e^{\frac{7800}{69167 RT}}\left(1.0197 p^{-1.8}\right)\right] \quad (7)$$

For calculating the start of combustion, the following equation is used [44]:

$$\int_{\theta_{SOI}}^{\theta_{SOI}+\Delta\theta_{IGD}} \frac{d\theta}{\Delta\theta_{IGD}(t)} = 1 \quad (8)$$

A constant value can be usually assumed for the combustion model parameter a, whereas the parameter m and the combustion duration $\Delta\theta$ depend on the engine speed and the air-fuel equivalence ratio according to the following equations:

$$\Delta\theta = \Delta\theta_o \left(\frac{\lambda_o}{\lambda}\right)^{k_1}\left(\frac{N}{N_o}\right)^{k_2} \quad (9)$$

$$m = m_o \left(\frac{\Delta\theta_{IGD,o}}{\Delta\theta_{IGD}}\right)^{k_1} \left(\frac{N_o}{N}\right)^{k_2} \frac{(pV/T)_{EVC}}{(pV/T)_{EVC,o}} \qquad (10)$$

where o denotes a reference point. It must be noted that the term $(pV/T)_{EVC}$ is proportional to the working medium mass trapped within the cylinder at the EVC point.

Typical values for the constants used in Equations (9) and (10) are provided in [46] for various diesel engine types including large diesel engines.

The Woschni model [48] was used for calculating the cylinder gas to wall heat transfer coefficient, according to the following equation:

$$h_{cyl} = k D^{-0.2} p^{0.8} T^{-0.55} w^{0.8} \qquad (11)$$

where h_{cyl} is the heat transfer coefficient, k is the model constant, D is the cylinder bore, p is the cylinder gas pressure, T is the cylinder gas temperature and w is a representative cylinder gas velocity. Constant temperature at cylinder walls was assumed in this work.

The heat rate transferred from the cylinder gas to the cylinder walls is calculated by using the following equation:

$$\dot{Q} = h_{cyl} \sum_j A_j \left(T_g - T_j\right) \qquad (12)$$

where j denotes piston, cylinder head, liner and exhaust valve, respectively.

For calculating the engine friction losses, an equation providing the engine friction mean effective pressure as a function of the cylinder maximum pressure and the average piston speed was used [49]. Subsequently, the cylinder torque due to friction is estimated by using friction mean effective pressure. The instantaneous cylinder torque is calculated by using the gross cylinder torque (calculated by using the cylinder indicated work) and cylinder torque due to friction. The cylinder torque is supplied in the crankshaft element, where it is used for calculating the crankshaft rotational speed.

In the two stroke marine engines, the scavenging process takes place from the opening of scavenging port (SPO) till the closing of exhaust valve (EVC). Therefore, two enabled subsystems in MATLAB/Simulink were developed and used: one for calculating the scavenging process, and the other for computing the rest cylinder working processes including compression, combustion, expansion and exhaust gas blowdown. The scavenging port and exhaust valve opening/closing timing is used for determining which enabled subsystem must be employed each time step.

The scavenging process is quite significant for the simulation of a two-stroke diesel engine considering that it accounts for a long period in the open cycle and it affects the trapped mass of the charge as well as its temperature. As a compromise between pure displacement and perfect mixing, a two-zone scavenging model is employed for the scavenging process simulation in this study [50]. The cylinder is divided into a fresh air zone and a residual gas zone. The mass flow rate of the fresh scavenging air entering to the residual gas zone is calculated according to the following equation by introducing the mixing factor k_{sca}:

$$\dot{m}_{12} = k_{sca}\dot{m}_a \qquad (13)$$

where $\dot{m}_a$ is the air mass flow rate entering the cylinder through scavenging port.

The mixing factor k_{sca}, which affects the cylinder temperature and gas composition at the end of open cycle (start of closed cycle), is calibrated at each load point. For deriving the required equations to model the scavenging process, the mass and energy conservation are considered in each one of the two zones, respectively. The burnt fuel fraction time derivative in residual gas zone is calculated using Equation (2) considering both $(\dot{m}\cdot\xi)_{in}$ and dm_{fb}/dt are zero, since zone 1 contains only fresh air and combustion does not take place. Equation (1) simplified accordingly to exclude the terms related to dissociation effects and combustion, was used to derive temperatures of zones 1 and 2. The required initial values were estimated by using the scavenging receiver temperature and the cylinder temperature before the start of the scavenging process (at SPO point) taken from the one zone enabled subsystem, respectively.

The mean temperature of the two zones, which is calculated by taking into account each zone temperature and the respective specific heat at constant volume, provides the initial value for the calculation of the temperature of one zone enabled subsystem at the next cycle. The cylinder pressure time derivative during the scavenging process was calculated according to the following equation that was derived by using the energy balance and the ideal gas law of each zone taking into account that the sum of each zone volume equals to the total cylinder volume and neglecting the effects of dissociation, (i.e., the values of $\partial u/\partial p$ and $\partial R/\partial p$ are considered to be equal to zero):

2. Modular Engine Model Description

$$\frac{dp}{dt} = p \frac{\sum_{i=1}^{2} \frac{V_i}{\gamma_i} \left[\frac{1}{c_{v,i} T_i} \left(B_i - \frac{\partial u_i}{\partial \varphi_i} \frac{d\varphi_i}{dt} \right) + \frac{1}{m_i} \frac{dm_i}{dt} + \frac{1}{R_i} \frac{dR_i}{dt} \right] - \frac{dV}{dt}}{\sum_{i=1}^{2} V_i / \gamma_i} \qquad (14)$$

where:

$$B_i = \frac{1}{m_i} \left(\sum_{sf} \dot{Q}_{i,sf} + \sum_j \dot{m}_{i,j} h_{i,j} - u_i \frac{dm_i}{dt} \right)$$

where i denotes the zones and j denotes the flows entering/exiting each zone.

The mass balance applied in each zone in conjunction with Equation (13) provides the following equations that are used for the calculation of each zone mass:

$$\frac{dm_1}{dt} = (1 - k_{sca}) \dot{m}_a \qquad (15)$$

$$\frac{dm_2}{dt} = k_{sca} \dot{m}_a - \dot{m}_e \qquad (16)$$

where $\dot{m}_e$ is the exhaust gas mass flow rate exiting through the cylinder exhaust valve.

In total, the two zone scavenging model employs six differential equations (Equation (1) appropriately simplified for each zone, mass balance for each zone (i.e., Equations (15) and (16)), Equation (2) for zone 2 and Equation (14)) in conjunction with the algebraic equations for calculating the working medium properties.

All cylinders of the diesel engine are regarded as identical. Therefore, the only parameter that differs in each cylinder is the phase angle. If malfunction or degradation of the engine cylinders is considered, the combustion and heat transfer as well as scavenging model parameters of the corresponding cylinder can be changed. The torque and inertia of each cylinder are summed up and provided as input parameters for the calculation of engine crankshaft rotational speed.

2.2.2. Scavenging Ports/Exhaust Valve

A quasi-steady one dimensional compressible flow consideration for orifices was adopted to calculate the mass flow rates through the scavenging ports and exhaust valves [44]. The valve/port elements use the instantaneous values for the valve/port equivalent area (derived by using the discharge coefficient and the geometric area) along with the pressure, temperature and properties of the working medium contained in the adjacent receivers. The valve/port mass flow rate is calculated as a function of the equivalence area, the pressure ratio and the properties of the working medium. The mass and energy flow rates are provided to the adjacent elements *i.e.*, the scavenging receiver and the cylinders for the scavenging ports; the cylinders and exhaust receiver for the exhaust valve.

2.2.3. Engine Receivers

The open thermodynamic system concept is also used for modelling the engine receivers. The mass conservation is applied for calculating the working medium mass in the receivers, whereas the temperature is derived by applying the energy balance according to Equation (1). As the effects of dissociation are neglected and the receiver volumes are constant, the terms $\partial u/\partial p$ and dV/dt used in Equation (1) are considered to be zero. Additionally, only fresh air is assumed to be contained in the scavenging receiver so that the term $d\varphi/dt$ is ignored, whereas the equivalence ratio and its derivative are taken into account for the respective calculations in the exhaust gas receiver. The receiver pressure is subsequently found by using the ideal gas law. No heat transfer was taken into account for the engine scavenging receiver, whereas the heat transferred from the exhaust gas to the ambient was calculated by considering the total heat transfer coefficient, the heat transfer area and the temperature difference.

2.2.4. Turbocharging System Components

The turbocharger compressor was modelled by using the non-dimensional parameters approach developed in the authors' previous work [41] that can accurately represent the compressor performance in the entire operating region including low speeds. The steps of the extension method taken from [41] are summarised below:

- Divide the compressor map into zones depending on the available constant speed curves.
- Digitize the provided compressor performance map.
- Calculate the non-dimensional parameters for the digitised compressor operating points.

- Derive the values of the parameters for each zone using curve fitting techniques.
- Derive the equations interrelating the actual maximum compressor efficiency with speed and the non-corrected maximum compressor efficiency with speed using curve fitting techniques.
- Provide all the above parameters as input to the compressor model.
- Calculate the corrected speed and identify the respective compressor map zone.
- Calculate the non-dimensional parameters.
- Calculate the non-dimensional flow coefficient for the respective zone.
- Derive the compressor volumetric flow rate.
- Calculate the non-dimensional torque coefficient for the respective zone.
- Calculate the compressor torque.
- Calculate the non-corrected compressor isentropic efficiency.
- Derive the corrected compressor isentropic efficiency.

The compressor block uses as input the turbocharger rotational speed (taken from the turbocharger shaft element) and compressor pressure ratio (estimated using the pressures of the adjacent blocks, the air cooler pressure drop, the air filter pressure drop and the blower pressure increase) and calculates the compressor efficiency and air mass flow. Subsequently, the rest compressor parameters including the compressor outlet temperature, the compressor torque and power are calculated.

The turbine block requires as input the digitised maps of the turbine efficiency and swallowing capacity. By using the pressure, temperature and properties of the turbine adjacent elements as well as the provided maps, the turbine efficiency and mass flow are calculated and subsequently used for the calculation of turbine torque, power and outlet temperature.

The pressure increase of each electric driven blower is considered to be polynomial function of each volumetric flow. The air temperature at the blower outlet is derived by considering the air temperature at the blower inlet, the pressure ratio as well as the blower efficiency by using the equation defining the blower isentropic efficiency [45].

The air cooler is modelled within the compressor block. The air cooler effectiveness is calculated by considering a quadratic function of air mass flow. The air temperature at the air cooler outlet is derived by using the air cooler effectiveness, the cooler water inlet temperature and the air temperature at the compressor outlet [45].

2.2.5. Shaft Elements

The shaft elements (crankshaft, turbocharger shaft) use the following angular momentum conservation equation to calculate their rotational speed:

$$\frac{dN}{dt} = \frac{30 \sum Q}{\pi I} \qquad (17)$$

where ΣQ is the sum of torques provided in the shaft element and I is the total inertia.

For the crankshaft element, the crank angle (at each time step) is derived by integrating the rotational speed and providing the initial crank angle value. The crank angle of each cylinder is calculated by using the respective phase angle.

2.2.6. Governor Element

A proportional-integral (PI) controller law is used for calculating the engine fuel rack position. Appropriate limiters are used to protect the engine during fast transient runs as proposed by the engine manufacturers.

2.2.7. Model Set up and Parameters Calibration

The steps required to set up a model for a given engine are as follows:

- Select the component blocks which sufficiently represent the engine layout from the models library.
- Connect the component blocks with the required connections.
- Insert the input data in all the blocks.
- Preliminary calibration of the model constants for a reference point and carrying out simulation runs.
- Fine tuning the model constants to obtain the required accuracy.

The following input data are needed to set up the model: the engine geometric data, the exhaust valve and scavenging port profiles, the steady state compressor and turbine performance maps, the constants of engine submodels (combustion and scavenging), the propeller loading and the ambient conditions. For integrating the time derivatives used in the model, initial values are also required for the following variables: the engine/propeller and turbocharger speeds, the temperature and pressure of the working medium contained in the engine cylinders and receivers as well as the gas composition for cylinders and exhaust receiver. When changing between the used subsystems (from the one-zone subsystem to the two-zones subsystem and *vice versa*), the required initial conditions were taken from the last time step of the previously enabled system.

3. RESULTS AND DISCUSSION

3.1. Validation of Zero-Dimensional Model

The operation of the MAN B&W 7K98MC two-stroke marine diesel engine was investigated using the described modular zero-dimensional engine model built in MATLAB/Simulink environment. The engine employs a constant pressure turbocharging system by using three turbocharger units connected in parallel to supply air to the engine scavenging receiver. The compressed air exiting the turbocharger compressors is forwarded to the air cooler units (one air cooler is installed downstream each compressor), which reduce the air temperature in order to increase its density. When the engine operates at low loads (lower than 40%), the turbocharger speed is too low and thus, the required air for the engine cylinders cannot be covered. Therefore, blowers driven by electric motors are used for providing the required air to the engine cylinders at the engine low load operating region. The blowers are connected between the air cooler and the scavenging receiver and are activated for low scavenging receiver pressure (typically less than 1.55 bar), whereas they are deactivated when scavenging receiver pressure exceeds 1.7 bar [51]. The engine details and the required input data were mainly extracted from the engine project guide [51], whilst the engine shop trial measurements were used for providing the data set used for the model validation. Table 1 contains the main engine parameters. The engine three turbocharger units as well as the installed air coolers and blowers were considered to have identical performance. The start of injection timing was estimated by using the engine variable injection timing (VIT) schedule. The engine was considered to operate on the propeller curve passing through the MCR point.

Table 1. MAN B&W 7K98MC engine parameters.

Bore	980 mm
Stroke	2660 mm
Number of cylinders	7
Brake power at MCR	40,100 kW
Engine speed at MCR	94 r/min
BMEP at MCR	18.2 bar
Turbocharger units	3

The combustion model was calibrated at 75% engine load, which was considered to be the reference point, in order to capture the cylinder combustion performance in the entire engine operating region (high and low loads). The required heat release rates were calculated by using the measured cylinder pressure diagrams. Based on this information, the parameters a, m and $\Delta\theta$ of Equation (5) were estimated for 75% engine load by using an optimisation algorithm considering an objective function with parameters the engine maximum pressure and brake specific fuel consumption. Subsequently, considering the pressure diagrams at other engine loads, the constants k_1 and k_2 used in Equations (9) and (10) were tuned to capture the combustion behaviour at the entire engine operating range. The actual (derived from measured pressure diagrams) and the calculated heat release rates at various engine loads are shown in Figure 2. As can be seen from Figure 2, the agreement between the model prediction and the measured values at 100% and 25% loads is considered acceptable for a single Vibe combustion model, as a compromise between the two extreme load points was achieved. Indeed, a lower value for the parameter m seems to be beneficial for improving the heat release rate shape at 100% load, however, it could make combustion prediction even worse at 25% load. Moreover, as it can be inferred from Figure 3 and Figure 4, there is already a satisfactory agreement with the maximum cylinder pressure for the investigated engine loads.

3. Results and Discussion

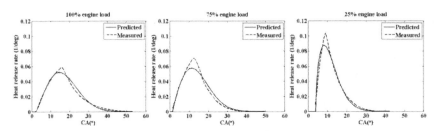

Figure 2. Heat release rates at various engine loads.

To enable the use of the model for investigating the engine operation with turbocharger cut-out, its ability to predict engine performance under normal engine operating conditions first needs to be examined. Therefore, the model was initially validated against the experimental data obtained from engine shop tests. The engine operation under steady state conditions at 25%, 50%, 75% and 100% of the MCR point was simulated. The obtained percentage error between the predicted engine performance parameters and the corresponding engine shop trial data are provided in Table 2. Acceptable accuracy is obtained for the entire engine operating region including 25% load, where the compressor operating point lays below the provided compressor map area. This shows that the used modelling approach, which includes the extrapolation of the compressor map towards the lowest speed area along with the blower and turbocharging system components models, is regarded as effective. It is deduced that the engine zero-dimensional model can provide adequate accuracy and therefore, it can be considered a reliable tool to simulate the examined engine operation cases presented below.

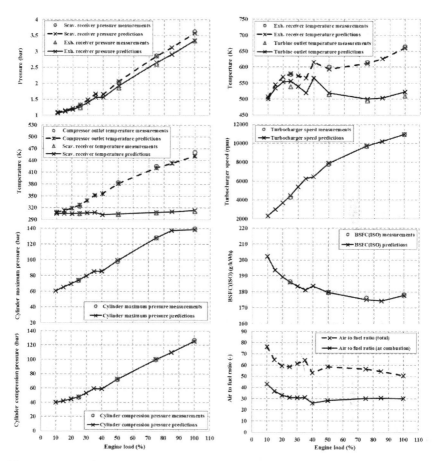

Figure 3. Steady state simulation results and comparison with shop trial data.

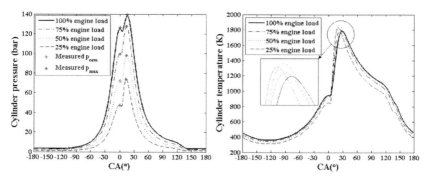

Figure 4. In-cylinder pressure and temperature variations at various loads.

3. Results and Discussion

Table 2. Steady state simulation results, comparison with shop trials data.

Engine Load (% MCR)	100	75	50	25
	\multicolumn{4}{c}{Error (%)}			
Brake power	−1.02	−1.01	−0.99	−0.92
BSFC	0.16	0.55	−0.08	0.18
Maximum cylinder pressure	0.71	0.07	−1.46	−0.59
Cylinder compression pressure	0.81	−0.23	−0.86	−1.16
Turbocharger speed	0.03	−0.41	−1.47	−4.52
Scavenging air receiver pressure	1.39	0.40	−0.89	−0.54
Exhaust gas receiver pressure	−0.20	−2.23	−4.11	−3.13
Scavenging air receiver temperature	−0.72	−0.39	−0.45	−0.24
Exhaust gas receiver temperature	0.46	0.48	1.08	−0.37
Exhaust gas temperature after turbocharger	2.52	−1.18	−0.99	−3.25

After the validation of the model, the engine operation at steady state conditions was investigated covering the entire load region from 10% to 100%. The activation of the electric driven blowers is induced for engine loads below 40% of the engine MCR, where the scavenging pressure becomes lower than 1.55 bar. The model predictions for a set of engine performance parameters including the receivers pressures and temperatures, the temperature of the exhaust gas exiting the turbine, the turbocharger speed, the brake specific fuel consumption (corrected at ISO conditions), the air to fuel ratio (total and at combustion), as well as the cylinder compression and maximum pressure are presented in Figure 3. The available engine recorded parameters for the engine loads 25%, 50%, 75% and 100% of MCR are also included in Figure 3. It is observed that the minimum brake specific fuel consumption is obtained at 85% load, at which the variable injection timing is the most advanced leading to almost the same maximum cylinder pressure level as 100% load.

Discontinuities in the engine performance parameters variations between 35% and 40% load can be observed, which are attributed to the engine blowers activation. Thus, a greater air amount enters the engine cylinders and therefore, the total and combustion air to fuel ratio values increase, whereas the temperatures of exhaust gas at turbine inlet (exhaust gas receiver temperature) and outlet become lower than the corresponding values at 40% load, where the blowers are switched off. The blower activation results in greater energy content provided to the turbine, which causes an increase in the turbocharger speed and scavenging receiver pressure in comparison with the respective values without blower activation (not shown in Figure 3).

This, in turn, increases the cylinder compression pressure and maximum pressure, thus, reducing the engine BSFC compared to the case without blower activation. Moreover, the blower compression process leads to an approximately 5 K increase of the temperature of the air contained in the scavenging receiver for loads lower than 35% (in comparison with the respective temperature value at 40% load). In the load range from 35% to 25% of MCR, the turbocharger compressor operates at lower speed delivering less air flow rate and as a result, the exhaust receiver temperature marginally increases. At engine loads 20% and lower, the exhaust receiver temperature decreases, since the injected fuel is more drastically reduced in comparison to the engine air flow (due to the turbocharger speed drop), and as a consequence, both the total and combustion air to fuel ratio values become greater.

The in-cylinder parameters *versus* crank angle are presented in Figure 4 showing a sufficient prediction accuracy of the maximum pressure and the compression pressure. As the engine load reduces, there is a reduction in the cylinder temperature during the open cycle. On the other hand, the maximum average cylinder temperature increases; at 75% load due to the earlier start of injection, whereas at 50% load due to the less trapped air amount, as it is inferred from the lower value of the air to fuel ratio at combustion. When the blowers are activated (below 40% load), the air to fuel ratio significantly increases leading to the reduction of the maximum cylinder temperature.

The scavenging process calculation used in this work is based on a two-zone model as explained in the previous sections. The results for the cylinder scavenging parameters variation during the exhaust gas blowdown and scavenging periods (from the opening of exhaust valve to the closing of exhaust valve) at 75% load are presented in Figure 5.

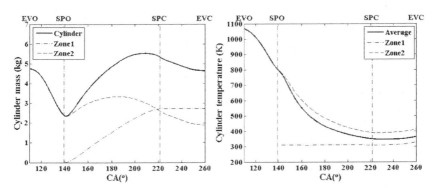

Figure 5. Scavenging model results for engine operation at 75% load.

3. Results and Discussion

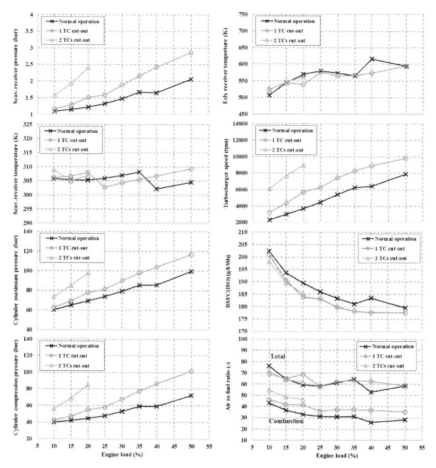

Figure 6. Steady state simulation results for normal operation and turbocharger cut-out. (Blowers are activated as follows: below 40% load for normal operation; below 25% load for operation with 1 TC cut-out; at 10% load for operation with 2 TCs cut-out).

The cylinder exhaust gas mass and temperature quickly reduce after the opening of the exhaust valve as the exhaust gas exits the cylinder. After the scavenging port opening (SPO), air enters into the cylinder and therefore, the air zone is created and grows. Since a part of the air passes from the air zone to the mixing zone, the mass of the latter increases after SPO and takes the maximum value at around 10 °CA after BDC. Afterwards, there is a reduction in the mixing zone mass as more exhaust gas exits the cylinder (entering the exhaust receiver). After the scavenging port closing (SPC) point, the exhaust gas continues exiting the cylinder and therefore, the trapped mass, which contains an amount of residual gas, is gradually reduced till the exhaust valve closing (EVC) point.

3.2. Engine Operation Investigation with Turbocharger Cut-Out

This section investigates the influence of turbocharger cut-out on the engine performance as well as the various alternatives for operating the engine at part loads down to 10% load, which imposes great challenges due to the significant change of the available exhaust gas flow area. Comparisons of the engine performance parameters predictions under normal operation with the respective ones obtained when using turbocharger cut-out for engine loads below 50% are presented in Figure 6.

The adoption of one turbocharger unit cut-out (out of three) caused a substantial rise of the turbocharger speed and as a result, the increase of the scavenging receiver pressure. In this respect, greater total engine air flow is obtained as it is deduced by the engine total air to fuel ratio values, which are greater than the ones of normal engine operation in the region from 40% to 50% (where the blowers are switched off) and almost the same with the ones of the normal engine operation for the region within 25% to 35% load (where the blowers are switched on). In addition, more air amount remains into the engine cylinders as it is observed by considering the combustion air to fuel ratio values, which increase in the load region from 25% to 50% of MCR. The increased scavenging receiver pressure results in the rise of the cylinder compression and maximum pressures and the consequential reduction of BSFC.

Due to the higher amount of air delivered by the compressor and trapped into the engine cylinders, the turbocharger cut-out decreases the exhaust receiver temperature and therefore, it lowers the thermal loading for engine load from 40% to 50%. It is also observed that the values of the total air to fuel ratio gradually reduced in the load range from 35% to 25% load with one turbocharger cut-out and it is estimated that the total air to fuel ratio will become lower than the respective one at normal operation below 25% load for the case where the engine operates without blower activation. This will result in a significant rise of exhaust receiver temperature imposing greater thermal loading to the engine components. Thus, it is inferred that the two operating turbochargers cannot provide adequate air amount to the engine cylinders. In consequence, the following solutions need to be considered for the engine operation below 25% load: either the switch on of the electric driven blowers or the cut-out of the second turbocharger unit. Both cases are also presented in Figure 6.

For the engine operation with one turbocharger cut-out at loads below 25%, the blower activation provides more air amount, increases the trapped air within cylinder (as can been inferred by the values of air to fuel ratio at combustion), and subsequently improves the engine BSFC. For operating the engine with two turbochargers cut-out, the increase of the turbocharger speed and the scavenging receiver pressure as well as the respective rise of the cylinder compression and maximum pressures are more noticeable resulting in an improvement of BSFC at 20% load comparing with the normal engine operation case.

The total air to fuel ratio value is comparable with the values observed for the normal operating cases, however, the air to fuel ratio at combustion is higher, which means that the trapped air amount is greater. Since two blowers should be activated at this operating point where one turbocharger is cut-out, the electrical energy required by them should also be taken into consideration for selecting the best alternative. Besides that, the engine is commonly not recommended to operate with blower activation for a long period. Therefore, it is deduced that two turbochargers cut-out (out of three) without blower activation is superior to one turbocharger cut-out with blower activation at 20% load.

The cylinder pressure differences (scavenging receiver pressure minus cylinder pressure, cylinder pressure minus exhaust receiver pressure and scavenging receiver pressure minus exhaust receiver pressure) for engine operation at 20%, 15% and 10% loads and for the three investigated cases (without turbocharger cut-out and activated blowers, with one turbocharger cut-out and with two turbochargers cut-out) are shown in Figure 7. The total engine air mass flow and the cylinders trapped air mass depend on the pressure differences, the scavenging ports and exhaust valve effective areas as well as the scavenging receiver air density (which mainly depends on the scavenging receiver pressure). Therefore, both parameters are affected by the number of turbochargers cut-out and the blowers activation. The cylinder pressure difference reduces as the number of cut-out turbochargers increases, since the total turbine geometric area reduces (this effect is equivalent as closing a variable geometry turbine area). However, this increases the scavenging receiver pressure and as a result, the air density; therefore the effect in the total engine air flow is limited and it is also affected by the blowers activation, as it can be inferred by the values of the total air to fuel ratio shown in Figure 6. On the other hand, the influence of the turbocharger cut-out on the cylinders air trapped amount is quite evident as it is deduced by the increased values of air to fuel ratio at combustion (Figure 6).

The cylinder pressure difference decreases as the engine load reduces, as shown in Figure 7, and it will become negative even when two turbochargers are cut-out in case where the blowers are kept deactivated below 15% load. For the two-stroke engines, the negative value of the cylinder pressure difference is not allowed as only the positive value ensures that fresh air will continue entering the engine cylinders; otherwise the engine cannot operate. In addition, the scavenging receiver pressure also drops below 1.55 bar, which is the blower activation pressure limit set by the engine manufacturer. Therefore, the blowers must be activated to supply the required air amount at such low engine loads (in the region of 10% load).

As can be seen from Figure 7, the cylinder pressure difference with one or two turbochargers cut-out and with the assistance of blower activation reaches the respective values observed for the normal operation. At 20% and 15% loads, there is a clear drop of the cylinder pressure difference with two turbochargers cut-out in comparison to its counterparts for the cases of the engine normal operation and the operation with one turbocharger cut-out and blower activation. That leads to less air amount into the cylinder in the scavenging process and the consequential higher cylinder temperature in the open cycle as shown in Figure 8 for all the examined cases of engine operation. At 10% load, the in-cylinder temperature variations at open cycle are comparable with each other, since the blowers are activated and they supply the required fresh air. The in-cylinder temperature in the closed cycle becomes lower as two turbochargers are cut-out due to the fact that more air is trapped within the cylinder at the end of the compression process, as it is also inferred from the comparison of the air to fuel ratio at combustion shown in Figure 6.

Based on the above investigation, the proposed engine part load operation schedule is as follows:

- From 50% to 25% load, one turbocharger cut-out without blower activation.
- From 20% to 15% load, two turbochargers cut-out without blower activation.
- In the region of 10% load, two turbochargers cut-out with blower activation.

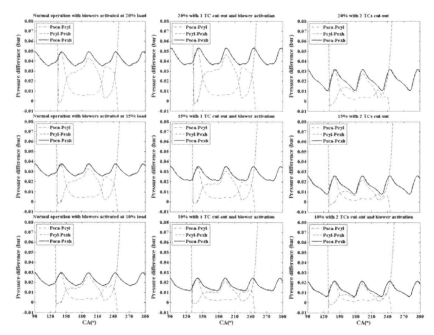

Figure 7. Cylinder pressure difference at different conditions.

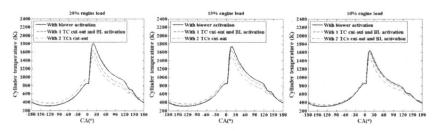

Figure 8. Cylinder temperature at different conditions.

According to [52], the investigated engine is generally used as the main propulsion engine in panamax type container vessels with capacity in the range of 4500 TEU. As it is reported in [53], the ship speed profile of this ship type has been varied from 2009 to 2012 towards lower ship speeds. The respective average ship speed value was changed from 22 knots at 2009 to around 14 knots at 2012. Considering that the engine power and ship speed in containerships is approximately in accordance to a power law with exponent values from 3.5 to 4.5 [54], the engine operating profile shown in Figure 9 was estimated and used for the calculation presented below. Based on this Figure, it is deduced that the ship main engine usually operates at loads lower than 50%, and as a result, increasing the engine efficiency at the low load region is crucial for reducing the ship operating cost.

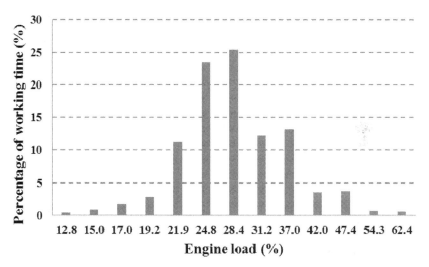

Figure 9. Engine operating profile of panamax vessels in 2012.

In order to analyse the fuel saving and CO_2 emissions reduction potential, the fuel consumption during the ship annual operation was calculated for the following three cases: (a) the engine normal operation; (b) the engine operation with one turbocharger cut-out at loads lower than 50%; and (c) the proposed herein engine part load operation schedule. In addition, the annual CO_2 emissions amount was estimated by using the CO_2 emission factor (3.114 kg CO_2/kg HFO) as proposed by IMO [55]. For the calculation presented below, it was assumed that the ship sails around 5900 h per year, which were estimated by excluding 10% and 25% of the total annual time for ship maintenance and port staying, respectively. The annual engine operating hours at each load were calculated by using the operating profile shown in Figure 9, whereas the engine brake specific fuel consumption was taken as function of the load and the operating mode from the data presented in Figure 6. The annual fuel consumption was calculated by using the following equation:

$$FC_a = H_a \sum_{i=1}^{n} \left(\frac{POT_i}{100} BSFC_{ISO,i} \frac{LHV_{ISO}}{LHV_{HFO}} P_{b,i} \right) \qquad (18)$$

where H_a is the engine annual operating hours, *POT* is the percentage of operating time at a specific load, $BSFC_{ISO}$ is the engine ISO corrected brake specific fuel consumption at this load, P_b is the engine brake power at this load, and *i* denotes the various engine loads according to the considered engine operating profile.

The calculation results for each operating strategy are presented in Figure 10. It must be noted that three blowers simultaneous operate under normal engine operating conditions; two blower units operate when one turbocharger unit is cut-out, whereas only one blower operates according the schedule recommended in this work. Since the same engine load profile was used in all the investigated cases, fuel savings are expected when the engine operates with one or two turbochargers cut-out due to the lower obtained engine BSFC as shown in Figure 6. Indeed, as it can be inferred from Figure 10, the engine operation with turbocharger cut-out can save annually up to 250 tonnes of fuel and reduce the CO_2 emissions by 760 tonnes per year, which corresponds to approximately 2% fuel and CO_2 emissions reduction compared with the normal operating mode.

It seems that the engine operation with one turbocharger cut-out and the proposed engine operation strategy can achieve a similar improvement in terms of the total fuel consumption and CO_2 emissions. However, the annual blowers operating time (per blower unit) is substantially reduced when the proposed turbocharger cut-out schedule is used as shown in Figure 11. This is expected to have a positive influence both to the maintenance cost and the electrical energy cost, as the electrical energy consumption needed for the operating blowers is very small for the case of the proposed turbocharger cut-out schedule due to the blowers limited operating hours.

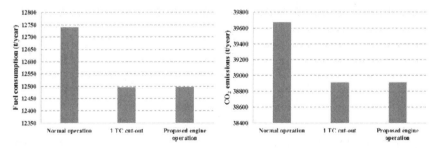

Figure 10. Calculated annual fuel consumption (**left**) and CO_2 emissions (**right**).

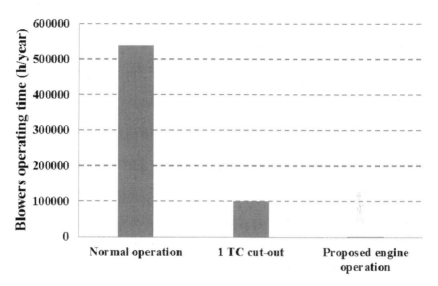

Figure 11. Annual blowers operating time (per operating unit); Normal operation: three blowers operate; Operation with 1 TC cut-out: two blowers operate; Proposed engine operation: one blower operates.

4. CONCLUSIONS

The study of two stroke marine diesel engine operation was carried out at steady state conditions focusing on part loads using a modular zero-dimensional model implemented in the MATLAB/Simulink environment. The main conclusions drawn from the present investigation are summarized as follows.

The developed modular engine model is a very effective way of engine modelling since the components submodels can be easily replaced based on the specific purpose of investigation. The used zero-dimensional model proved its capability of adequately predicting the engine steady state performance including the in-cylinder parameters variation in the whole engine operating envelope down to the ultra-low load region with high accuracy. The model complexity is greater comparing with the MVEM alternative, but it is compensated by the model enhanced predictive ability.

For the case of engine normal operation at loads lower than 50% of MCR, the cylinder maximum pressure reduces drastically as scavenging receiver pressure decreases, leading to greater brake specific fuel consumption. In addition, the exhaust receiver temperature increases resulting in a higher thermal loading as engine load reduces since the engine air to fuel ratio becomes lower. As a consequence, the blowers activation is required at low load region ensuring a reliable engine operation. The blowers activation results in a significant increase of a number of engine parameters including the turbochargers speed, the scavenging pressure and the engine air amount. Therefore, the cylinder maximum pressure and the air to fuel ratio increase, which in turn, improves the engine brake specific fuel consumption and decreases the exhaust gas temperature, resulting in lower thermal loading of the engine and its components. As an alternative option, the turbocharger cut-out can achieve equivalent or slightly better results with regards to the fuel consumption. Based on the conducted analysis, the preferable engine operation scenario at part loads was found to be as follows: one turbocharger cut-out without blower activation is preferred for the engine operating region between 25% and 50% load; two turbochargers cut-out without blower activation is favoured when the engine operates between 15% and 20% load; two turbochargers cut-out with blower activation should be used at around 10% load.

The proposed turbocharger cut-out schedule can reduce the annual fuel consumption of a panamax containership operating at slow steaming conditions by approximately 2% with a similar reduction in the released CO_2 emissions compared with the respective operation at slow steaming conditions without turbocharger cut-out. In addition, it can reduce the operating hours of the electric driven blowers and the associated electrical energy consumption. Therefore, a positive influence on the ship operating cost from both fuel and maintenance as well as the environment is expected.

ACKNOWLEDGMENTS

The research work performed by Guan and Professor Chen was supported by the Programme of Introducing Talents of Discipline to Universities (Grant No. B08031), the Research Fund for the Doctoral Program of Higher Education of China (Grant No. 20100143110004), and the Natural Science Foundation of Hubei Province of China (Grant No. 2014CFB843).

AUTHOR CONTRIBUTIONS

Cong Guan and Gerasimos Theotokatos contributed in equal parts to the establishment of the model, the calculation and the writing, whereby Hui Chen had some useful suggestions and been involved in the discussion and preparation of the manuscript.

REFERENCES

1. Carolina, G.M.; Julio, R.; Maria, J.S. Modelling and forecasting fossil fuels, CO_2 and electricity prices and their volatilities. *Appl. Energy* **2013**, *101*, 363–375.
2. Zhang, C.; Chen, X. The impact of global oil price shocks in China's bulk commodity markets and fundamental industries. *Energy Policy* **2014**, *66*, 32–41.
3. Zhang, Y.; Wang, Z. Investigating the price discovery and risk transfer functions in the crude oil and gasoline futures markets: Some empirical evidence. *Appl. Energy* **2013**, *104*, 220–228.
4. *Alphaliner: Container Ship Charter Rates to Weaken in 2013*, Available online: https://www.joc.com/comment/1601(accessed on 5 December 2014).
5. Mandavia, M.; Bureau, E.T. *Shipping industry may confront another deluge of overcapacity*, Available online: http://articles.economictimes.indiatimes.com/2014-04-02/news/48801359_1_great-eastern-shipping-essar-shipping-shipping-industry (accessed on 5 December 2014).
6. *Analysis: Global Shipping Contends with Oversupply Problems*, Available online: http://www.stratfor.com/analysis/global-shipping-contends-oversupply-problems#axzz3L2lNkuZT (accessed on 5 December 2014).
7. Gcaptain. *BIMCO Reflections: Oversupply Will Delay Shipping Market Recovery*, Available online: http://gcaptain.com/bimco-reflections-2014-shipping-industry-analysis/ (accessed on 5 December 2014).
8. International Maritime Organisation (IMO). *Air Pollution and Energy Efficiency-Estimated CO_2 Emissions Reduction from Introduction of Mandatory Technical and Operational Energy*

Efficiency Measures for Ships; IMO: London, UK, 2011; MEPC 63/INF.2.

9. Haglind, F. A review on the use of gas and steam turbine combined cycles as prime movers for large ships. Part III: Fuels Emiss. *Energy Convers. Manag.* **2008**, *49*, 3476–3482.

10. Pappas, C.; Karakosta, C.; Marinakis, V.; Psarras, J. A comparison of electricity production technologies in terms of sustainable development. *Energy Convers. Manag.* **2012**, *64*, 626–632.

11. Wärtsilä. *Marine Solutions*, 2nd ed.; Wärtsilä: Helsinki, Finland, 2012; Publication no: SP-EN-DBAC136254.

12. MAN Diesel & Turbo. *Marine Engine IMO Tier II Programme 2013*; MAN Diesel & Turbo: Augsburg, Germany, 2013; Publication no. 4510-0012-00ppr.

13. Brown, D. Helping shipowners cut fuel bills with Wärtsilä low-speed engines. *Wärtsilä Tech. J. Mar./InDetail* **2009**, *1*, 34–37.

14. MAN Diesel & Turbo. *SFOC Optimization Methods for MAN B&W Two-Stroke IMO Tier II Engines*; MAN Diesel & Turbo: Augsburg, Germany, 2012; Publication no. 5510-0099-00ppr.

15. MAN Diesel & Turbo. *Waste Heat Recovery System (WHRS) for Reduction of Fuel Consumption, Emissions and EEDI*; MAN Diesel & Turbo: Augsburg, Germany, 2012; Publication no. 5510-0136-01ppr.

16. Grljušić, M.; Medica, V.; Račić, N. Thermodynamic analysis of a ship power plant operating with waste heat recovery through combined heat and power production. *Energies* **2014**, *7*, 7368–7394.

17. Schmid, H. Less emissions through waste heat recovery. In Proceedings of the Green Ship Technology Conference, London, UK, 28–29 April 2004; Wärtsilä Corporation: Helsinki, Finland.

18. MAN Diesel & Turbo. *Slide Fuel Valve*; MAN Diesel & Turbo: Augsburg, Germany, 2010; Publication no. 1510-0011-03ppr.

19. MAN Diesel & Turbo. *Operation on Low-Sulphur Fuels*; MAN Diesel & Turbo: Augsburg, Germany, 2014; Publication no. 5510-0075-01ppr.

20. ABS. *Ship Energy Efficiency Measures*; ABS: Houston, TX, USA, 2013; Publication no: TX 05/13 5000-13015.

21. Armstrong, V.N. Vessel optimisation for low carbon shipping. *Ocean Eng.* **2013**, *73*, 195–207.

22. Wiesmann, A. Slow steaming-a viable long-term option? *Wärtsilä Tech. J. Mar./InDetail* **2010**, *2*, 49–55.
23. Pierre, C. Is slow steaming a sustainable means of reducing CO_2 emissions from container shipping? *Transp. Res. Part D* **2011**, *16*, 260–264.
24. Alphaliner: Slow steaming absorbs capacity. Available online: http://www.transportjournal.com/en/home/news/artikeldetail/slow-steaming-absorbs-capacity.html (accessed on 5 December 2014).
25. Gard, A.S. Slow steaming on 2-stroke engines. Available online: http://www.gard.no/ikbViewer/Content/8259/No%2003-09%20Slow%20Steaming%20on%202%20stroke%20engines.pdf (accessed on 5 December 2014).
26. Schmuttermair, H.; Fernandez, A.; Witt, M. Fuel economy by load profile optimized charging systems from MAN. In Proceedings of the 26th CIMAC World Congress on Combustion Engine Technology, Bergen, Norway, 14–17 June 2010. Paper no. 250.
27. Baechi, R. Slow steaming and turbocharger cut-out. *ABB Charge* **2012**, *2*, 20–21.
28. Woodward, J.B.; Latorre, R.G. Modeling of diesel engine transient behaviour in marine propulsion analysis. *Trans. Soc. Nav. Archit. Mar. Eng.* **1984**, *92*, 33–49.
29. Hendricks, E. Mean Value Modelling of Large Turbocharged Two-Stroke Diesel Engines, SAE Technical Paper no 890564. 1989.
30. Chesse, P.; Chalet, D.; Tauzia, X. Real-time performance simulation of marine diesel engines for the training of navy crews. *Mar. Technol.* **2004**, *41*, 95–101.
31. Theotokatos, G. On the cycle mean value modelling of a large two-stroke marine diesel engine. *Proc. IMechE Part M: J. Eng. Mar. Environ.* **2010**, *224*, 193–205.
32. Katrašnik, T. Transient momentum balance-A method for improving the performance of mean-value engine plant models. *Energies* **2013**, *6*, 2892–2926.
33. Kyrtatos, N.P.; Koumbarelis, I. Performance prediction of next-generation slow speed diesel engines during ship manoeuvers. *Trans. IMarE* **1994**, *106*, 1–26.

34. Finesso, R.; Spessa, E. A real time zero-dimensional diagnostic model for the calculation of in-cylinder temperatures, HRR and nitrogen oxides in diesel engines. *Energy Convers. Manag.* **2014**, *79*, 498–510.

35. Catania, A.E.; Finesso, R.; Spessa, E. Predictive zero-dimensional combustion model for DI diesel engine feed-forward control. *Energy Convers. Manag.* **2011**, *52*, 3159–3175.

36. Asad, U.; Tjong, J.; Zheng, M. Exhaust gas recirculation–zero dimensional modelling and characterization for transient diesel combustion control. *Energy Convers. Manag.* **2014**, *86*, 309–324.

37. Chmela, F.G.; Pirker, G.H.; Wimmer, A. Zero-dimensional ROHR simulation for DI diesel engines–A generic approach. *Energy Convers. Manag.* **2007**, *48*, 2942–2950.

38. Xiros, N. *Robust Control of Diesel Ship Propulsion*; Springer-Verlag London Ltd.: London, UK, 2002.

39. Theotokatos, G.; Tzelepis, V. A computational study on the performance and emission parameters mapping of a ship propulsion system. *Proc. IMechE Part M: J. Eng. Mar. Environ.* **2015**, *229*, 58–76.

40. Livanos, G.A.; Papalambrou, G.; Kyrtatos, N.P.; Christou, A. Electronic engine control for ice operation of tankers. In Proceedings of the 25th CIMAC World Congress on combustion Engine Technology, Vienna, Austria, 21–24 May 2007. Paper no. 44.

41. Guan, C.; Theotokatos, G.; Zhou, P.; Chen, H. Computational investigation of a large containership propulsion engine operation at slow steaming conditions. *Appl. Energy* **2014**, *130*, 370–383.

42. Hountalas, D.T.; Sakellaridis, N.F.; Pariotis, E.; Antonopoulos, A.K.; Zissimatos, L.; Papadakis, N. Effect of turbocharger cut out on two-stroke marine diesel engine performance and NOx emissions at part load operation. In Proceedings of the ASME 12th biennial conference on engineering systems design and analysis, Copenhagen, Denmark, 25–27 July 2014. ESDA2014-20514.

43. Kyrtatos, N.; Glaros, S.; Tzanos, E.; Hatzigrigoris, S.; Dalmyras, F. Systematic evaluation of performance of VLCC engine, comparing service monitored data and thermodynamic model predictions. In Proceedings of the 27th CIMAC World Congress on

Combustion Engine Technology, Shanghai, China, 13–16 May 2013. Paper no. 32.

44. Heywood, J.B. *Internal Combustion Engines Fundamentals*; McGraw-Hill: New York, USA, 1988.

45. Watson, N.; Janota, M.S. *Turbocharging the Internal Combustion Engine*; MacMillan: London, UK, 1982.

46. Merker, G.P.; Schwarz, C.; Stiesch, G.; Otto, F. *Simulating Combustion*; Springer-Verlag: Berlin, Germany, 2006.

47. Klein, M. Single-Zone Cylinder Pressure Modeling and Estimation for Heat Release Analysis of SI Engines. Ph.D. Dissertation, Linköping University, Linköping, Sweden, 2007.

48. Woschni, G. A Universally Applicable Equation for the Instantaneous Heat Transfer Coefficient in the Internal Combustion Engine, SAE Paper no 670931. 1967.

49. Ciulli, E. A Review of Internal Combustion Engine Losses. Part 2: Studies for Global Evaluations. *Proc. Inst. Mech. Eng.* **1992**, *207*, 229–240. Merker, G.; Gerstle, M. Evaluation on Two Stroke Engines Scavenging Models, SAE Technical Paper no 970358. 1997.

50. MAN Diesel & Turbo. *MAN B&W K98MC Project Guide: Two-Stroke Engines*, 3rd ed.; MAN Diesel & Turbo: Augsburg, Germany, 2002.

51. MAN Diesel & Turbo. *Propulsion Trends in Container Vessels*; MAN Diesel & Turbo: Augsburg, Germany, 2013; No. 5510-0040-02ppr.

52. Banks, C.; Turan, O.; Incecik, A.; Theotokatos, G.; Izkan, S.; Shewell, C.; Tian, X. Understanding ship operating profiles with an aim to improve energy efficient ship operations. In Proceedings of the Low Carbon Shipping Conference, London, UK, 9–10 September 2013.

53. MAN Diesel & Turbo. *Basic Principles of Ship Propulsion*; MAN Diesel & Turbo: Augsburg, Germany, 2011; No. 5510-0004-02ppr.

54. International Maritime Organisation (IMO). *Third IMO GHG Study*; IMO: London, UK, 2014; MEPC 67/INF.3.

Performance Analysis of the Vehicle Diesel Engine-ORC Combined System Based on a Screw Expander

Kai Yang [1], Hongguang Zhang [1,*], Songsong Song [1,2], Jian Zhang [1], Yuting Wu [1], Yeqiang Zhang [1], Hongjin Wang [1], Ying Chang [1] and Chen Bei [1]

[1] College of Environmental and Energy Engineering, Beijing University of Technology, Pingleyuan No.100, Beijing 100124, China

[2] Automotive Engineering Department, Chengde Petroleum College, Chengde, Hebei 067000, China

ABSTRACT

To achieve energy saving and emission reduction for vehicle diesel engines, the organic Rankine cycle (ORC) was employed to recover waste heat from vehicle diesel engines, R245fa was used as ORC working fluid, and the resulting vehicle diesel engine-ORC combined system was presented. The variation law of engine exhaust energy rate under various operating conditions was obtained, and the running performances of the screw expander were introduced. Based on thermodynamic models and theoretical calculations, the running performance of the vehicle diesel engine-ORC combined system was analyzed under various engine operating condition scenarios. Four evaluation indexes were defined: engine thermal efficiency increasing ratio (ETEIR), waste heat recovery efficiency (WHRE), brake specific fuel consumption (BSFC) of the combined system, and improvement ratio of BSFC (IRBSFC). Results showed that when the diesel engine speed is 2200 r/min and diesel engine torque is 1200 N·m, the power output of the combined system reaches its maximum of approximately 308.6 kW, which is 28.6 kW higher than that of the diesel engine. ETEIR, WHRE, and IRBSFC all reach their maxima at 10.25%, 9.90%, and 9.30%, respectively. Compared with that of the diesel engine, the BSFC of the combined system is obviously improved under various engine operating conditions.

Keywords: waste heat recovery; vehicle diesel engine; organic Rankine cycle; screw expander; various operating conditions

1. INTRODUCTION

Internal combustion (IC) engines consume a large amount of petroleum resources. The thermal efficiency of IC engines is less than 40%. A large proportion of the energy from fuel combustion is released in the form of waste heat into the atmosphere through the exhaust and the coolant system [1,2]. Waste heat recovery is an effective means to improve fuel consumption, save energy, and reduce IC engine emissions [3].

The organic Rankine cycle (ORC) system is considered effective in converting low-grade waste heat to useful work and has recently been widely studied and applied in many domains [4–7]. Wang et al. [8] established an off-design model of an ORC system driven by solar energy. El-Emam et al. [9] presented thermodynamic and economic analyses on a novel type of geothermal regenerative ORC system. Uris et al. [10] assessed the technical and economic feasibility of biomass-fueled ORC power plants. Carcasci et al. [11] indicated that the use of an ORC is a promising choice for the recovery of waste heat at low or medium temperatures.

Many researchers have concluded that the ORC system is a highly effective means of recovering waste heat for IC engines [12–15], thus the topic has become a research hot spot worldwide. Peris et al. [16] utilized the ORC system to recover the waste heat of jacket cooling water from IC engines. Meinel et al. [17] recovered the exhaust energy of IC engines by means of an ORC system. Hajabdollahi et al. [18] built a model of an ORC for diesel engine waste heat recovery and analyzed the thermal efficiency and the total annual cost of the system.

In an ORC system, the match of organic working fluids with heat source and systems significantly affects system performance. Numerous researchers have conducted studies on organic working fluid selection [19–22]. Wang et al. [23] analyzed nine different pure organic working fluids and indicated that R245fa and R245ca are the most suitable working fluids for engine waste heat recovery applications. Lakew et al. [24] concluded that R245fa can provide high power output for temperatures higher than 160 °C. Rayegan *et al.* [25] asserted that R245fa and R245ca are suitable working fluids for an ORC system at medium temperature. Based on the literature survey, R245fa performs suitably as the working fluid in an ORC system because of good thermodynamic and environmental performance.

As a key component of the ORC system, an expander is used to produce useful work, and the running performance of the expander has a crucial effect on the running performance of the ORC system, including such aspects as net power output (W_{net} [26–30]. Kang *et al.* [31] conducted an ORC capable of generating electric power with a radial turbine and analyzed the influence factors of the ORC system. Twomey *et al.* [32] tested the performance of a scroll expander in a small ORC system. Qiu *et al.* [33] concluded that vane expanders and scroll expanders might be the best choices for micro-scale combined heat and power systems.

In practice, a vehicle IC engine generally runs under various operating conditions, and the amount of waste heat from IC engine varies with these engine operating conditions. To recover the waste heat efficiently and effectively under engine various operating conditions, it is crucial to know the variation law of engine exhaust energy and select an ideal expander. In this paper, by experiment, the variation law of engine exhaust energy rate under various operating conditions was obtained, and the running performances of the screw expander were investigated, then the vehicle diesel engine-ORC combined system was designed. Furthermore, by theoretical calculation, the running performances of the vehicle diesel engine-ORC combined system were analyzed under various operating conditions of the engine.

As we all know, there are many electrical equipments powered by the electricity generator and battery in the vehicle. Generally, the electricity generator is driven by the vehicle engine, which certainly decreases the net power output of the vehicle engine. In this research, an ORC system is adopted to recover waste heat from diesel engine exhaust, and the screw expander used in the ORC system, in place of vehicle engine, is employed to drive the electricity generator. In this way, we can effectively improve the fuel consumption of the vehicle engine. Furthermore, electricity generation based on waste heat recovery of vehicle engine may also be an efficient way of saving energy and reducing emissions for the internal combustion engine–electric motor hybrid vehicle in the near future.

2. EXPERIMENTAL STUDY ON EXHAUST ENERGY RATE AND SCREW EXPANDER

2.1. Available Exhaust Energy Rate of Vehicle Diesel Engine

The IC engine used for the study of waste heat recovery is a six-cylinder and four-stroke vehicle diesel engine. The main parameters of the vehicle diesel engine are listed in Table 1. The diesel engine experimental system is illustrated in Figure 1. The test was performed under engine various operating conditions, including more than 85 operating condition points. During the diesel engine test, the engine speed varied from 600 r/min to 2200 r/min, and the engine torque varied from 0 N·m to 1500 N·m. Some of the tested operating condition points are listed in Table 2.

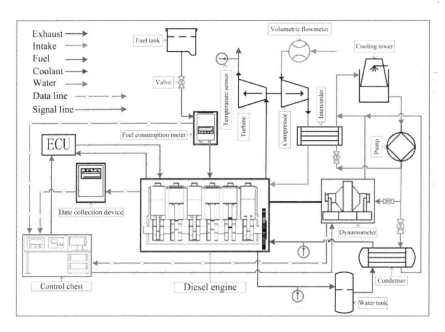

Figure 1. Schematic diagram of the diesel engine experimental system.

Table 1. Main parameters of the diesel engine.

Item	Parameter	Unit
Displacement	9726	mL
Cylinder diameter	126	mm
Stroke	130	mm
Rated speed	2200	r/min
Maximum torque	1500	N·m
Rated power	280	kW
Compression ratio	17	-

Table 2. Tested engine operating condition points.

Engine speed (r/min)	Engine torque (N·m)	Engine power (kW)	Fuel consumption rate (kg/h)	Intake air flow rate (kg/h)	Engine exhaust temperature (K)
2200	1214.9	279.87	66.01	1661.6	818.95
2000	1313.9	275.17	60.97	1554.2	783.45
1800	1420.8	267.8	56.7	1432.4	779.85
1600	1514	253.67	50.76	1252.4	745.95
1400	1543.2	226.24	44.14	1090.1	710.75
1200	1537.3	193.18	36.83	918.4	704.05
1000	1183.6	123.95	23.18	595.4	650.75
800	1001	83.86	16.38	361.7	653.95
600	610.9	38.38	8.24	214.7	568.85

The variation trend of brake specific fuel consumption (BSFC) of the diesel engine under various operating conditions is shown in Figure 2. When the engine speed is lower than 1100 r/min, BSFC gradually decreases with the increase of engine torque. When the engine speed is higher than 1100 r/min, BSFC gradually decreases initially and then gradually increases with the increase of engine torque. When the engine torque is lower than 400 N·m, BSFC gradually decreases initially and then gradually increases with the increase of engine speed. When the engine runs with high speed and low torque, BSFC is relatively high. When engine speed is 1100 r/min and engine torque is 1300 N·m, BSFC reaches its minimum.

2. Experimental Study on Exhaust Energy Rate and Screw Expander

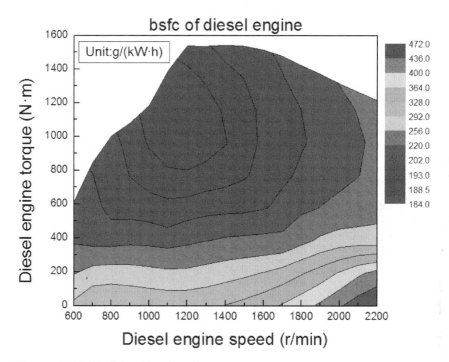

Figure 2. BSFC of the diesel engine.

The available exhaust energy rate ($\dot{Q}_{ava}$) of the diesel engine can be calculated as follows:

$$\dot{Q}_{ava} = c_p \dot{m}_{exh} (T_{exh\_1} - T_{min}) \quad (1)$$

where, $\dot{m}_{exh}$ is the exhaust mass flow rate of the diesel engine and is the sum of the intake air flow rate and fuel consumption rate ($\dot{m}_{fuel}$), which can be obtained during the diesel engine test; T_{exh_1} is the engine exhaust temperature at the inlet of the evaporator of the engine exhaust side and can be obtained during the diesel engine test; T_{min} is the available minimum temperature of the exhaust at the outlet of the evaporator of the engine exhaust side and is set to 303.15 K; and c_p is the isobaric specific heat of engine exhaust, which can be calculated as follows:

$$c_P = 0.00025 T_{exh\_1} + 0.99 \qquad (2)$$

The variation law of available exhaust energy rate under various operating conditions is shown in Figure 3. The engine available exhaust energy rate gradually increases with the increase of engine torque and engine speed, and the maximum of engine available exhaust energy rate is 290.0 kW. From Table 1, it is shown that the rated power of the diesel engine is 280 kW, which is lower than the maximum of engine available exhaust energy rate, so it is meaningful to recover and utilize the waste heat of the diesel engine exhaust.

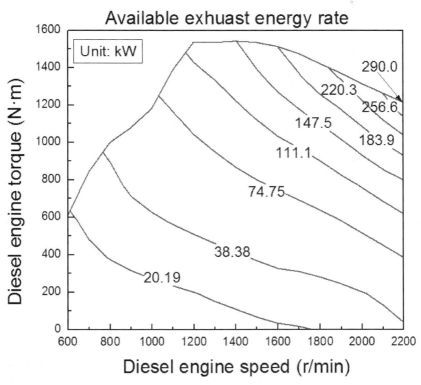

Figure 3. Available exhaust energy rate of the diesel engine.

2.2. Screw Expander

The screw expander experimental system is illustrated in Figure 4. The screw expander was designed and manufactured by our research group at the Beijing University of Technology [34-37]. The screw expander experimental system mainly consists of screw expander, working fluid (compressed air) circuit, lubricating oil circuit, power testing system, data acquisition system and water cooling system. The experiment uses compressed air as working fluid, ambient air is sucked into the compressor and pressurized, and finally be discharged to the ambient environment after the expansion process in the screw expander. The air flow rate entering the screw expander can be adjusted through the regulation valve at the outlet of the gasholder. The lubricating oil is driven by an oil pump and flows into the screw expander for the purpose of lubricating and sealing. An oil separator is used to remove lubricating oil in the air which flows out of the screw expander. An eddy current dynamometer is used to measure the power produced by the screw expander, and the water cooling system is employed to cool down the eddy current dynamometer. Different parameters, such as flow rate, inlet and outlet pressure, inlet and outlet temperature, rotational speed, torque, and power, are measured.

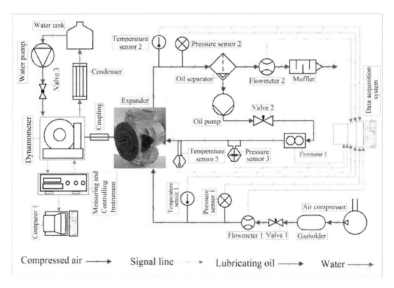

Figure 4. Schematic diagram of the screw expander experimental system.

The variation of screw expander power output with the inlet pressure and rotational speed of the screw expander is shown in Figure 5. With the increase of inlet pressure and rotational speed of screw expander, the power output of the screw expander gradually increases. When the inlet pressure is 1.7 MPa and rotational speed is 3200 r/min, the power output reaches its maximum of approximately 51.20 kW.

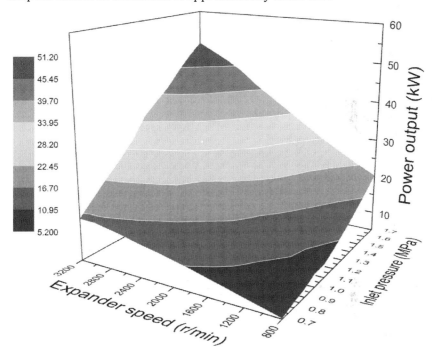

Figure 5. Variation of power output with inlet pressure and rotational speed.

The expansion ratio is the ratio of inlet pressure to outlet pressure of the screw expander. The variation tendency of the expansion ratio with inlet pressure is shown in Figure 6, which demonstrates that the expansion ratio evidently varies with the inlet pressure and rotational speed. Overall, the expansion ratio tends to become lower with the increase of rotational speed. Furthermore, when the inlet pressure is relatively low, the expansion ratio tends to become higher with the increase of inlet pressure. When the inlet pressure is 1.7 MPa and rotational speed is 3200 r/min, the expansion ratio of the screw expander is approximately 8.

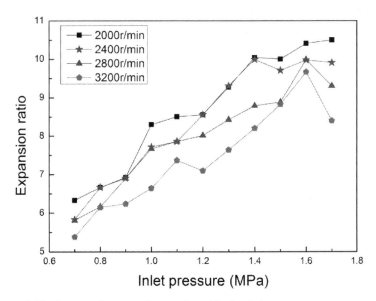

Figure 6. Variation of expansion ratio with the inlet pressure.

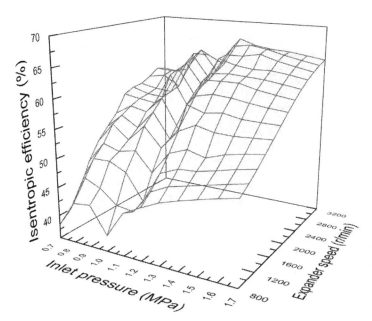

Figure 7. Variation of isentropic efficiency with inlet pressure and rotational speed.

The variation of screw expander isentropic efficiency with the inlet pressure and rotational speed of the screw expander is shown in Figure 7. It can be seen that the isentropic efficiency evidently varies with the inlet pressure and rotational speed. Overall, isentropic efficiency tends to increase with the increase of rotational speed. Furthermore, when the inlet pressure is relatively high, isentropic efficiency tends to become higher with the increase of inlet pressure. When the inlet pressure is 1.7 MPa and rotational speed is 3200 r/min, the isentropic efficiency of the screw expander is approximately 0.65.

3. VEHICLE DIESEL ENGINE-ORC COMBINED SYSTEM

3.1. Configuration of the Combined System

In this paper, on the basis of studying variation law of engine exhaust energy rate and running performances of the screw expander, the vehicle diesel engine-ORC combined system is presented, the combined system mainly consists two parts: vehicle diesel engine, ORC system. The working process of the vehicle diesel engine is based on the diesel cycle, whereas the working process of the ORC system is based on the Rankine cycle. For the combined system, diesel cycle is the topping cycle, and Rankine cycle is the bottoming cycle. The vehicle diesel engine-ORC combined system is illustrated in Figure 8. The ORC system mainly consists of an evaporator, screw expander, condenser, reservoir, and pump. When the ORC system is running, the working fluid is drawn from the reservoir and pressurized into a subcooled liquid state by the pump. The working fluid is then sent to the evaporator and is heated by engine exhaust. The working fluid turns into saturated vapor state with high temperature and high pressure. The saturated vapor flows into the screw expander to produce useful work. After the expansion process, with a decrease in temperature and pressure, the superheated vapor exhausted from the screw expander enters the condenser, and condenses into saturated liquid state in the condenser, then flows into the reservoir. The whole organic Rankine cycle (ORC) process is completed. R245fa is used as the working fluid for the ORC system, and its main properties are listed in Table 3.

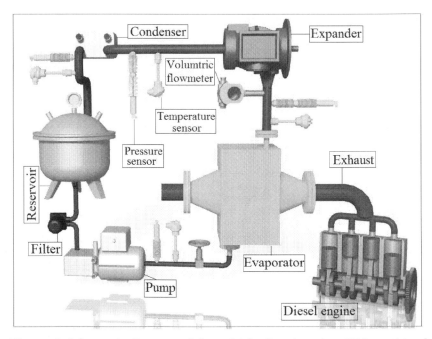

Figure 8. Schematic diagram of the vehicle diesel engine-ORC combined system.

Table 3. Main properties of the working fluid R245fa.

Working fluid	Chemical formula	Molar mass (kg/kmol)	$T_{critical}$ (K)
R245fa	$CHF_2CH_2CF_3$	134.05	427.16
$P_{critical}$ (MPa)	$\rho_{critical}$ (kg/m^3)	ODP	GWP (100 years)
3.651	516.08	0.0	950

The evaporator employed for this research is a finned-tube heat exchanger, which has a larger heat transfer area to improve the heat transfer rate between the working fluid and engine exhaust compared to a shell-and-tube heat exchanger. Moreover, a finned-tube heat exchanger has lower flow resistance. Initially, a plate heat exchanger was used as the condenser. Through some preliminary experiments, we found that the plate heat exchanger has higher flow resistance, which results in higher outlet pressure of the expander and lower net power output of the ORC system. In the next stage, we plan to use a finned-tube heat exchanger as the condenser in order to optimize the running performance of the ORC system and the combined system. At present, a multistage centrifugal pump is selected as the working fluid pump due to its merits such as stable operation, low vibration and long working life. By regulating valves installed in the pipelines, the mass flow rate of the working fluid can be effectively adjusted.

The single screw expander has many advantages, such as balanced loading of the main screw, long working life, high volumetric efficiency, high expansion ratio, low noise, low vibration and compact configuration, *etc*. Quite a few kinds of fluids, such as high pressure gas, superheated steam, saturated steam, gas-liquid two-phase fluid and hot liquid can be used as the working fluid for single screw expander.

3.2. Thermodynamic Model

The *T-s* diagram of the ORC system is shown in Figure 9, where T_{exh_1} is the engine exhaust temperature at the inlet of the evaporator of the engine exhaust side, T_{exh_3} is the engine exhaust temperature at the outlet of the evaporator of the engine exhaust side, Process $T_{exh\_1} - T_{exh\_3}$ is the heat rejection process of the engine exhaust in the evaporator, and ΔT_{pp} is the pinch point temperature difference (PPTD) between the diesel engine exhaust and the working fluid R245fa, and it is set to 10 K in this paper. Process 1-2 is the actual expansion process of the working fluid in the screw expander. Process 1-2s is the isentropic expansion process. Process 2-3 is the isobaric condensing process of the working fluid in the condenser. Process 3-4 is the actual compression process of the working fluid in the pump. Process 3-4s is the isentropic compression process. Process 4-1 is the isobaric endothermic process of the working fluid in the evaporator. Because the engine exhaust temperature varies with engine operating condition, according to the different operating conditions of the diesel engine, occurrence position of pinch point temperature difference (PPTD) between the diesel engine exhaust and the working fluid R245fa may change, which may appear at the inlet of the evaporator of the working fluid side (state point 4 in Figure 9), at the outlet of the evaporator of the working fluid side (state point 1 in Figure 9), or at the saturated liquid state point of the working fluid (state point a in Figure 9). On the basis of our previous research, we can conclude that when the diesel engine torque is higher than 300 N·m, the occurrence position of pinch point temperature difference (PPTD) between the engine exhaust and working fluid certainly appears at state point 4 in Figure 9 (the inlet of the evaporator of the working fluid side). The main thermodynamic parameters of each state point of the ORC system are listed in Table 4.

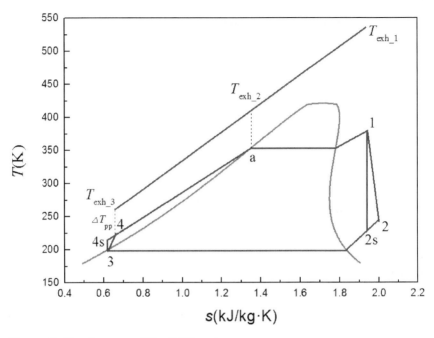

Figure 9. *T-s* diagram of the ORC system.

Table 4. Thermodynamic parameters of each state point of the ORC system.

Parameters	State point 1	State point 2	State point 2s	State point 3	State point 4	State point 4s
Temperature [K]	386.973	335.251	321.653	308.258	309.027	308.815
Pressure [MPa]	1.700	0.213	0.213	0.213	1.700	1.700
Enthalpy [kJ·kg^{-1}]	481.637	456.767	443.376	245.952	247.369	247.086
Entropy [kJ·(kg·K)$^{-1}$]	1.797	1.838	1.797	1.157	1.158	1.157

The power output of the screw expander can be expressed as:

$$\dot{W}_s = \dot{m}(h_1 - h_2) = \dot{m}(h_1 - h_{2s})\eta_s \quad (3)$$

The heat transfer rate between hot fluid and cold fluid in the condenser can be expressed as:

$$\dot{W}_p = \dot{m}(h_4 - h_3) = \frac{\dot{m}(h_{4s} - h_3)}{\eta_p} \qquad (4)$$

The power consumption of the pump can be expressed as:

$$\dot{W}_p = \dot{m}(h_4 - h_3) = \frac{\dot{m}(h_{4s} - h_3)}{\eta_p} \qquad (5)$$

The heat transfer rate between hot fluid and cold fluid in the evaporator can be expressed as:

$$\dot{Q}_e = \dot{m}(h_1 - h_4) \qquad (6)$$

The net power output of the ORC system can be calculated as follow:

$$\dot{W}_{net} = \dot{W}_s - \dot{W}_p \qquad (7)$$

The power output of the combined system ($\dot{W}_{com}$) can be calculated as follow:

$$\dot{W}_{com} = \dot{W}_{net} + \dot{W}_{ice} \qquad (8)$$

To objectively evaluate the running performances of the combined system, four evaluation indexes of engine thermal efficiency increasing ratio (ETEIR), waste heat recovery efficiency (WHRE), brake specific fuel consumption (BSFC) of the combined system, and improvement ratio of BSFC (IRBSFC) are proposed.

ETEIR can be calculated as follows:

$$\text{ETEIR} = \frac{(\dot{W}_{ice} + \dot{W}_{net})/\dot{Q}_{fuel} - \dot{W}_{ice}/\dot{Q}_{fuel}}{\dot{W}_{ice}/\dot{Q}_{fuel}} = \frac{\dot{W}_{net}}{\dot{W}_{ice}} \quad (9)$$

WHRE can be calculated as follows:

$$\text{WHRE} = \frac{\dot{W}_{net}}{\dot{Q}_{ava}} \quad (10)$$

BSFC of the combined system (bsfc$_{com}$) can be calculated as follows:

$$\text{bsfc}_{com} = \frac{\dot{m}_{fuel}}{\dot{W}_{ice} + \dot{W}_{net}} \quad (11)$$

IRBSFC (η_{bsfc}) can be calculated as follows:

$$\eta_{bsfc} = \frac{\dot{m}_{fuel}/\dot{W}_{ice} - \dot{m}_{fuel}/(\dot{W}_{ice} + \dot{W}_{net})}{\dot{m}_{fuel}/\dot{W}_{ice}} \quad (12)$$

From the aforementioned experimental results, it can be seen that, when the inlet pressure is 1.7 MPa and rotational speed is 3200 r/min, the expansion ratio of the screw expander is approximately 8, and the isentropic efficiency of the screw expander is approximately 0.65. Furthermore, the power output of the screw expander reaches its maximum. Thus, the above-mentioned parameter values are selected for the design operating point of the screw expander in the combined system, which indicates that the evaporating pressure of the ORC system can be set to 1.7 MPa. Moreover, for the ORC system, pressure drop and heat loss of components and pipelines are neglected, isentropic efficiency of the pump is set to 0.8, pinch point temperature difference between diesel engine exhaust and working fluid R245fa (ΔT_{pp}) is set to 10 K.

4. CALCULATION RESULTS AND DISCUSSION

The power output of the vehicle diesel engine-ORC combined system under engine various operating conditions is shown in Figure 10. It can be concluded that, power output of the combined system gradually increases with the increase of engine speed and engine torque. The main reason for this is, with the increase of engine speed and engine torque, both power out of the diesel engine and net power output of the ORC system increase. When diesel engine speed is 2200 r/min and diesel engine torque is 1200 N·m, the power output of the combined system reaches its maximum of approximately 308.6 kW. The rated power of the diesel engine is 280 kW, and the power output of the combined system is 28.6 kW higher than that of the diesel engine.

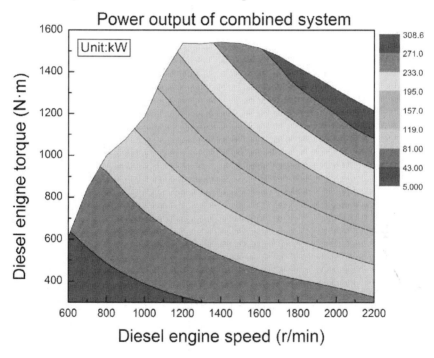

Figure 10. Power output of the combined system.

The mass flow rate variation of the working fluid R245fa under engine various operating conditions is shown in Figure 11. It can be concluded that, with the increase of engine speed and engine torque, the mass flow rate of the working fluid gradually increases. This condition can be attributed to the fact that with the increase of engine speed and engine torque, engine available exhaust energy rate increases, such that more working fluid can be heated and evaporated in the evaporator. When diesel engine speed is 2200 r/min and diesel engine torque is 1200 N·m, the mass flow rate of the working fluid reaches its maximum of approximately 1.22 kg/s. The required mass flow rate of the working fluid varies with the operating condition of the diesel engine. Considering the variation of engine available exhaust energy rate, the mass flow rate of the working fluid should be actively regulated for the corresponding operating condition of the diesel engine, which is helpful for recovering the engine exhaust energy rate efficiently and effectively. Figures 10 and 11 indicate that the distribution tendency of the contour lines of R245fa mass flow rate is similar to that of the contour lines of the combined system power output. This condition indicates that mass flow rate has an important effect on power output of the combined system under engine various operating conditions.

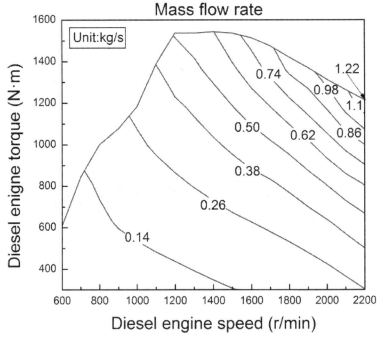

Figure 11. Mass flow rate of the working fluid R245fa.

Under various operating conditions of the engine, the power output of the combined system is higher than that of the diesel engine. To assess the improvement of fuel economy and thermal efficiency, the concept of ETEIR is presented. The variation trend of ETEIR under engine various operating conditions is shown in Figure 12. When the engine torque is certain (in this paper, "certain" means "be held constant", the same as below), ETEIR gradually increases with the increase of engine speed. When the engine speed is in the range of 600 r/min to 1200 r/min,

ETEIR gradually decreases with the increase of engine torque. When the engine speed is in the range of 1200 r/min to 2200 r/min, with the increase of engine torque, ETEIR decreases initially and then increases. It can be seen that, when the diesel engine runs with high speed, ETEIR is relatively high. This observation can be attributed to several factors. First, when the diesel engine runs with high speed and low torque, the power output of the diesel engine is relatively low, and engine available exhaust energy rate is relatively high. Thus, the net power output of the ORC system is relatively high and ETEIR becomes higher. Second, when the diesel engine runs with high speed and high torque, engine available exhaust energy rate is higher, and net power output of the ORC system is higher, then ETEIR is higher. When diesel engine speed is 2200 r/min and the diesel engine torque is 1200 N·m, ETEIR reaches its maximum of approximately 10.25%.

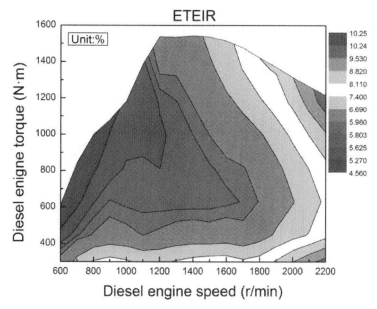

Figure 12. ETEIR under engine various operating conditions.

In order to assess the utilization ratio of engine available exhaust energy rate, WHRE is defined. The variation tendency of WHRE under engine various operating conditions is shown in Figure 13. When the engine speed is certain, WHRE gradually increases with the increase of engine torque. When the engine torque is in the range of 300 N·m to 400 N·m, WHRE decreases initially and then increases with the increase of engine speed. When the engine torque is in the range of 400 N·m to 1200 N·m,

4. Calculation Results and Discussion

WHRE increases initially, then decreases, and increases anew with the increase of engine speed. When the engine torque is higher than 1200 N·m, WHRE gradually increases with the increase of engine speed. This result is mainly ascribed to the fact that both engine available exhaust energy rate and net power output of the ORC system may vary with the operating condition of the diesel engine; moreover, the amplitude of variation of the engine available exhaust energy rate may be different from the amplitude of variation of the ORC system power output. When the engine runs with high speed and high torque, WHRE is relatively high, the maximum of WHRE is approximately 9.9%.

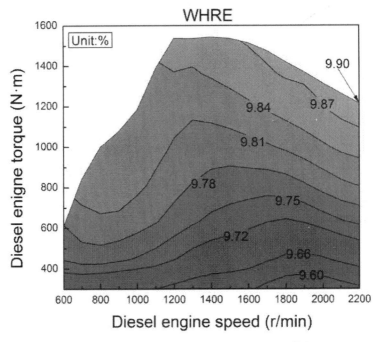

Figure 13. WHRE under engine various operating conditions.

The variation tendency of the BSFC of the combined system under engine various operating conditions is shown in Figure 14. When the engine torque is in the range of 300 N·m to 1350 N·m, BSFC of the combined system decreases initially and then increases with the increase of engine speed. When the engine torque is higher than 1350 N·m, BSFC of the combined system gradually increases with the increase of engine speed. When the engine speed is in the range of 600 r/min to 1100 r/min, BSFC of the combined system gradually decreases with the increase of engine torque. When the engine speed is in the range of 1100 r/min to 2200 r/min, BSFC of the combined system decreases initially and then increases with the increase of engine torque. Figures 2 and 14 indicate that the variation tendency of the BSFC of the combined system is similar to the variation tendency of BSFC of the diesel engine. According to the same engine operating condition, the BSFC of the combined system is lower than that of the diesel engine.

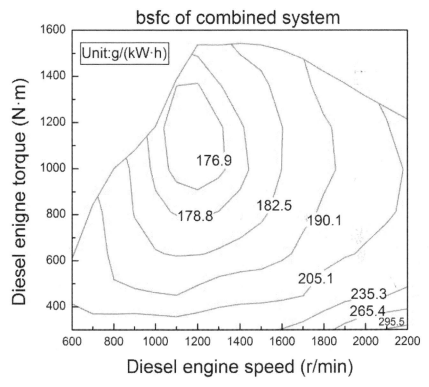

Figure 14. BSFC of the combined system under engine various operating conditions.

IRBSFC is proposed to assess the improvement of BSFC between the combined system and the diesel engine. The variation trend of the IRBSFC under engine various operating conditions is shown in Figure 15. When the engine torque is higher than 900 N·m, IRBSFC gradually increases with the increase of engine speed. When the engine speed is in the range of 600 r/min to 1100 r/min, IRBSFC gradually decreases with the increase of engine torque. When the engine speed is in the range of

1100 r/min to 2200 r/min, IRBSFC decreases initially and then increases with the increase of engine torque. This observation is attributed to the fact that both the BSFC of the diesel engine and the net power output of the ORC system may vary with the operating condition of the diesel engine; moreover, the amplitude of variation of the diesel engine BSFC may be different from the amplitude of variation of the ORC system power output.

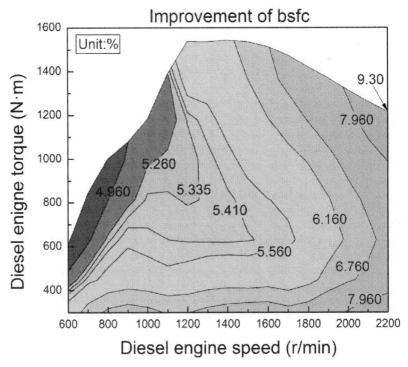

Figure 15. IRBSFC under engine various operating conditions.

5. CONCLUSIONS

In this paper, a vehicle diesel engine-ORC combined system was designed, and the variation law of the engine exhaust energy rate, running performances of the screw expander, and running performances of the vehicle diesel engine-ORC combined system were studied.

1. Key parameter values were determined experimentally for the design operating point of the screw expander in the combined system. The parameter values are as follows: the evaporating pressure of the ORC system is 1.7 MPa, rotational speed of the screw expander is 3200 r/min, expansion ratio of the screw expander is 8, and isentropic efficiency of the screw expander is 0.65.

2. With the increase of engine speed and engine torque, the power output of the combined system gradually increases. When diesel engine speed is 2200 r/min and diesel engine torque is 1200 N·m, the power output of the combined system reaches its maximum of 308.6 kW. The rated power of the diesel engine is 280 kW, and the power output of the combined system is 28.6 kW higher than that of the diesel engine.

3. According to the same engine operating conditions, the BSFC of the combined system is lower than that of the diesel engine. When diesel engine speed is 2200 r/min and diesel engine torque is 1200 N·m, ETEIR, WHRE, and IRBSFC all reach their maxima, which are 10.25%, 9.90%, and 9.30%, respectively.

4. The engine available exhaust energy rate varies with engine operating conditions. In order to recover the engine exhaust energy efficiently, the mass flow rate of the working fluid should be actively regulated for the corresponding operating conditions of the diesel engine.

ACKNOWLEDGMENTS

This work was sponsored by the National Natural Science Foundation of China (Grant No. 51376011); Scientific Research Key Program of Beijing Municipal Commission of Education (Grant No. KZ201410005003); The National Basic Research Program of China (973 Program) (Grant No. 2011CB707202); and The Twelfth Scientific Research Foundation for Graduate Students in Beijing University of Technology (Grant No. ykj-2013-9386).

REFERENCES

1. Shu, G.; Zhao, J.; Tian, H.; Wei, H.Q.; Liang, X.Y.; Yu, G.P.; Liu, L.N. Theoretical analysis of engine waste heat recovery by the combined thermo-generator and organic Rankine cycle system. Proceedings of SAE 2012 World Congress & Exhibition, Detroit, MI, USA, 24–26 April 2012.

2. Bari, S.; Hossain, S.N. Waste heat recovery from a diesel engine using shell and tube heat exchanger. *Appl. Therm. Eng.* **2013**, *61*, 355–363.

3. Yun, K.T.; Cho, H.; Luck, R.; Mago, P.J. Modeling of reciprocating internal combustion engines for power generation and heat recovery. *Appl. Energy* **2013**, *102*, 327–335.

4. Zhang, J.H.; Feng, J.C.; Zhou, Y.L.; Fang, F.; Yue, H. Linear active disturbance rejection control of waste heat recovery systems with organic rankine cycles. *Energies* **2012**, *5*, 5111–5125.

5. Reverberi, A.; Borghi, A.D.; Dovì, V. Optimal design of cogeneration systems in industrial plants combined with district heating/cooling and underground thermal energy storage. *Energies* **2011**, *4*, 2151–2165.

6. Kaska, O. Energy and exergy analysis of an organic Rankine for power generation from waste heat recovery in steel industry. *Energy Convers. Manag.* **2014**, *77*, 108–117.

7. Karellas, S.; Leontaritis, A.D.; Panousis, G.; Bellos, E.; Kakaras, E. Energetic and exergetic analysis of waste heat recovery systems in the cement industry. *Energy* **2013**, *58*, 147–156.

8. Wang, J.F.; Yan, Z.Q.; Zhao, P.; Dai, Y.P. Off-design performance analysis of a solar-powered organic Rankine cycle. *Energy Convers. Manag.* **2014**, *80*, 150–157.

9. El-Emam, R.S.; Dincer, I. Exergy and exergoeconomic analyses and optimization of geothermal organic Rankine cycle. *Appl. Therm. Eng.* **2013**, *59*, 435–444.

10. Uris, M.; Linares, J.I.; Arenas, E. Techno-economic feasibility assessment of a biomass cogeneration plant based on an Organic Rankine Cycle. *Renew. Energy* **2014**, *66*, 707–713.

11. Carcasci, C.; Ferraro, R.; Miliotti, E. Thermodynamic analysis of an organic Rankine cycle for waste heat recovery from gas turbines. *Energy* **2014**, *65*, 91–100.

12. Boretti, A.; Osman, A.; Aris, I. Design of Rankine cycle systems to deliver fuel economy benefits over cold start driving cycles. Proceedings of SAE 2012 International Powertrains, Fuels & Lubricants Meeting, Malmo, Sweden, 18-20 September 2012.

13. Glavatskaya, Y.L.; Podevin, P.; Lemort, V.; Shonda, O.; Descombes, G. Reciprocating Expander for an Exhaust Heat Recovery Rankine Cycle for a Passenger Car Application. *Energies* **2012**, *5*, 1751-1765.

14. Wei, M.S.; Fang, J.L.; Ma, C.C.; Danish Syed, N. Waste heat recovery from heavy-duty diesel engine exhaust gases by medium temperature ORC system. *Sci. China Technol. Sci.* **2011**, *54*, 2746-2753.

15. Xie, H.; Yang, C. Dynamic behavior of Rankine cycle system for waste heat recovery of heavy duty diesel engines under driving cycle. *Appl. Energy* **2013**, *112*, 130-141.

16. Peris, B.; Navarro-Esbrí, J.; Molés, F. Bottoming organic Rankine cycle configurations to increase Internal Combustion Engines power output from cooling water waste heat recovery. *Appl. Therm. Eng.* **2013**, *61*, 364-371.

17. Meinel, D.; Wieland, C.; Spliethoff, H. Effect and comparison of different working fluids on a two-stage organic Rankine cycle (ORC) concept. *Appl. Therm. Eng.* **2014**, *63*, 246-253.

18. Hajabdollahi, Z.; Hajabdollahi, F.; Tehrani, M.; Hajabdollahi, H. Thermo-economic environmental optimization of Organic Rankine Cycle for diesel waste heat recovery. *Energy* **2013**, *63*, 142-151.

19. Latz, G.; Andersson, S.; Munch, K. Comparison of working fluids in both subcritical and supercritical Rankine cycles for waste-heat recovery systems in heavy-duty vehicles. Proceedings of SAE 2012 World Congress & Exhibition, Detroit, MI, USA, 24-26 April 2012.

20. Gao, H.; Liu, C.; He, C.; Xu, X.X.; Wu, S.Y.; Li, Y.R. Performance analysis and working fluid selection of a supercritical organic Rankine cycle for low grade waste heat recovery. *Energies* **2012**, *5*, 3233-3247.

21. Liu, C.; He, C.; Gao, H.; Xu, X.X.; Xu, J.L. The optimal evaporation temperature of subcritical ORC based on second law efficiency for waste heat recovery. *Entropy* **2012**, *14*, 491-504.

22. Roy, J.P.; Misra, A. Parametric optimization and performance analysis of a regenerative Organic Rankine Cycle using R-123 for waste heat recovery. *Energy* **2012**, *39*, 227–235.

23. Wang, E.H.; Zhang, H.G.; Fan, B.Y.; Ouyang, M.G.; Zhao, Y.; Mu, Q.H. Study of working fluid selection of organic Rankine cycle (ORC) for engine waste heat recovery. *Energy* **2011**, *36*, 3406–3418.

24. Lakew, A.A.; Bolland, O. Working fluids for low-temperature heat source. *Appl. Therm. Eng.* **2010**, *30*, 1262–1268.

25. Rayegan, R.; Tao, Y.X. A procedure to select working fluids for Solar Organic Rankine Cycles (ORCs). *Renew. Energy* **2011**, *36*, 659–670.

26. Yagoub, W.; Doherty, P.; Riffat, S.B. Solar energy-gas driven micro-CHP system for an office building. *Appl. Therm. Eng.* **2006**, *26*, 1604–1610.

27. Manolakos, D.; Papadakis, G.; Kyritsis, S.; Bouzianas, K. Experimental evaluation of an autonomous low-temperature solar Rankine cycle system for reverse osmosis desalination. *Desalination* **2007**, *203*, 366–374.

28. Declaye, S.; Quoilin, S.; Guillaume, L.; Lemort, V. Experimental study on an open-drive scroll expander integrated into an ORC (Organic Rankine Cycle) system with R245fa as working fluid. *Energy* **2013**, *55*, 173–183.

29. Yamada, N.; Minami, T.; Mohamad, M.N.A. Fundamental experiment of pumpless Rankine-type cycle for low-temperature heat recovery. *Energy* **2011**, *36*, 1010–1017.

30. Qiu, G.Q.; Shao, Y.J.; Li, J.X.; Liu, H.; Riffat, S.B. Experimental investigation of a biomass-fired ORC-based micro-CHP for domestic applications. *Fuel* **2012**, *96*, 374–382.

31. Kang, S.H. Design and experimental study of ORC (organic Rankine cycle) and radial turbine using R245fa working fluid. *Energy* **2012**, *41*, 514–524.

32. Twomey, B.; Jacobs, P.A. Gurgenci, H. Dynamic performance estimation of small-scale solar cogeneration with an organic Rankine cycle using a scroll expander. *Appl. Therm. Eng.* **2013**, *51*, 1307–1316.

33. Qiu, G.Q.; Liu, H.; Riffat, S.B. Expander for micro-CHP systems with organic Rankine cycle. *Appl. Therm. Eng.* **2011**, *31*, 3301–3307.

34. Wang, W.; Wu, Y.T.; Ma, C.F.; Liu, L.D.; Yu, J. Preliminary experimental study of single screw expander prototype.*Appl. Therm. Eng.* **2011**, *31*, 3684–3688.

35. Wang, W.; Wu, Y.T.; Ma, C.F.; Xia, G.D.; Wang, J.F. Experimental study on the performance of single screw expanders by gap adjustment. *Energy* **2013**, *62*, 379–384.

36. He, W.; Wu, Y.T.; Peng, Y.H.; Zhang, Y.Q.; Ma, C.F.; Ma, G.Y. Influence of intake pressure on the performance of single screw expander working with compressed air. *Appl. Therm. Eng.* **2013**, *51*, 662–669.

37. Lu, Y.W.; He, W.; Wu, Y.T.; Ji, W.N.; Ma, C.F.; Guo, H. Performance study on compressed air refrigeration system based on single screw expander. *Energy* **2013**, *55*, 762–768.

CHAPTER 8

Combustion and Exhaust Emission Characteristics of Diesel Micro-Pilot Ignited Dual-Fuel Engine

Ulugbek Azimov[1], Eiji Tomita[2] and Nobuyuki Kawahara[2]

[1] Department of Mechanical Engineering, Curtin University, Malaysia campus,

[2] Department of Mechanical Engineering, Okayama University, Japan

1. INTRODUCTION

To satisfy increasingly strict emissions regulations, engines with alternative gaseous fuels are now widely used. Natural gas and synthesis gas appear to be greener alternatives for internal combustion engines [1-3]. In many situations where the price of petroleum fuels is high or where supplies are unreliable, the syngas, for example, can provide an economically viable solution. Syngas is produced by gasifying a solid fuel feedstock such as coal or biomass. The biomass gasification means incomplete combustion of biomass resulting in production of combustible gases. Syngas consists of about 40% combustible gases, mainly carbon monoxide (CO), hydrogen (H_2) and methane (CH_4). The rest are non-combustible gases and consists mainly of nitrogen (N_2) and carbon dioxide (CO_2). Varying proportions of CO_2, H_2O, N_2, and CH_4 may be present [4].

H_2 as a main component of a syngas has very clean burning characteristics, a high flame propagation speed and wide flammability limits. H_2 has a laminar combustion speed about eight times greater than that of natural gas, providing a reduction of combustion duration and as a result, an increase in the efficiency of internal combustion (IC) engines, if the H_2 content in the gaseous fuel increases. Main point of interest in increasing

1. Introduction

H_2 content in the gaseous fuel is that with the addition of H_2, the lean limit of the gas operation can be extended, without going into the lean misfire region. Lean mixture combustion has a great potential to achieve higher thermal efficiency and lower emissions [5]. In particular, the lean mixture combustion will result in low and even extremely low NOx levels with only a slight increase in hydrocarbons [6, 7].

Some gas engines fueled with syngas have been developed recently [8-10]. Most of them have a spark-ignition (SI) combustion system. An SI engine is not suitable for this kind of fuel under high load conditions because of the difficulty in achieving stable combustion due to the fluctuation of the syngas components. In addition, the syngas is a low energy density fuel and the extent of power degrading is large when compared with high-energy density fuels like gasoline and natural gas. Natural gas and syngas have high auto-ignition temperature and hence cannot be used in CI engines without a means of initiating combustion, as the temperature attained at the end of the compression stroke is too low for the mixture to be auto-ignited. Therefore, dual-fuel-mode engine operation is required, in which gaseous fuel is ignited by pilot diesel fuel. Dual fueling can serve as a way of allowing the current fleet of CI engines to reduce their dependence on conventional diesel fuel while minimizing harmful emissions.

A number of researchers have performed experiments with natural gas and syngas to determine engine performance and exhaust emissions in dual-fuel engines. Their results indicate that lower NOx and smoke can be achieved in dual-fuel engines compared with conventional diesel engines, while maintaining the same thermal efficiency as a diesel engine. McTaggart-Cowan *et al.* [11] investigated the effect of high-pressure injection on a pilot-ignited, directly injected natural gas engine. They found that at high loads, higher injection pressures substantially reduce PM emissions. At low loads, the amount of PM emissions are independent of the injection pressure. Without EGR, NOx emissions are slightly increased at higher injection pressures due to the faster and more intense combustion caused by improved mixing of air and fuel and increased in-cylinder temperature. Su and Lin [12] studied the amount of pilot injection and the rich and lean boundaries of natural gas dual-fuel engines. They found that there is a critical amount of pilot diesel fuel for each load and speed. Tomita *et al.* [13] investigated the combustion characteristics and performance of the supercharged syngas with micro-pilot (injected fuel - 2 mg/cycle) ignition in a dual-fuel engine. They found that premixed flame of syngas-air mixture develops from multiple flame kernels produced by the ignition of diesel fuel. It was found that with the certain increase of hydrogen content in syngas the engine could operate even at equivalence ratio of 0.45 with stable combustion and high efficiency, because the increased hydrogen content enhanced the lean limit of the mixture. Liu and

Karim [14] concluded that the observed values of the ignition delay in dual-fuel operation are strongly dependent on the type of gaseous fuels used and their concentrations in the cylinder charge. They showed that changes in the charge temperature during compression, preignition energy release, external heat transfer to the surroundings, and the contribution of residual gases appear to be the main factors responsible for controlling the length of the ignition delay of the engine.

The autoignition of the premixed mixture in the end-gas region is affected by the composition of syngas, in particular, by the amount of H_2, CO, CO_2 and CH_4 in the gas. H_2 has low ignition energy, and therefore, is easier to ignite, that results in a stronger tendency to autoignition and knock. H_2 has a flame speed much greater than that of hydrocarbon fuels. Also, it has a lean limit of ϕlim=0.1, much lower than the theoretical limit of methane (ϕlim=0.5) [15]. Carbon dioxide, on the other hand, can weaken the reactivity of the in-cylinder mixture by diluting it, which results in a longer ignition delay time and slower heat release rate. Methane has excellent anti-knock properties, but suffers from low flame propagation rates and high auto-ignition temperature. Carbon monoxide can also affect the reactivity of the mixture. In fact, the oxidation of CO in the presence of H_2 is important question concerning the syngas oxidation mechanism. It is well known that the overall reactivity is greatly accelerated if trace amounts of H_2 and moisture are present. The oxidation route between CO and OH is the dominant pathway, and it accounts for a significant portion of the heat release [16].

The aforementioned results suggest that if certain operating conditions are maintained, including the control of pilot fuel injection quantity, pressure and timing, gaseous fuel equivalence ratio, and EGR rate, a compromise between increased efficiency and low exhaust emissions can be achieved. In this chapter, we document the range of operating conditions under which the new higher-efficiency PREMIER (PREmixed Mixture Ignition in the End-gas Region) combustion mode was experimentally tested. The objective of this work was in brief to discuss conventional micro-pilot injected dual-fuel combustion and in detail to explain about PREMIER combustion and emission characteristics in a pilot ignited supercharged dual-fuel engine fueled with natural gas and syngas, and to study the effect of H_2 and CO_2 content in syngas on the combustion and emission formation over the broad range of equivalence ratios under lean conditions.

2. CONVENTIONAL MICRO-PILOT INJECTED DUAL-FUEL COMBUSTION

2.1. Experimental Procedure and Conditions

This study used water-cooled four-stroke single-cylinder engine, with two intake and two exhaust valves, shown in Figure 1 (A). In this engine, the autoignition of a small quantity of diesel pilot fuel (2 mg/cycle), injected into the combustion chamber before top dead center, initiates the combustion. The burning diesel fuel then ignites the gaseous fuel. The pilot fuel was ultra low-sulfur (<10ppm) diesel. A commercial solenoid-type injector that is typically used for diesel-only operations was modified. A nozzle of the commercial injector with seven holes was replaced by the one with four holes of 0.1 mm in diameter to ensure a small quantity of injected fuel.

Diesel fuel injection timing and injection duration were controlled through the signals transferred to the injector from the injector driver. A common rail injection system (ECD U2-P, Denso Co.) was employed to supply the constant injection pressure to the injector. The common rail pressure was set and controlled via computer. The fuel injection pressure varied from 40 MPa to 150 MPa, and the injected pilot diesel fuel quantity was 2 mg/cycle and 3 mg/cycle. The experimental conditions and different types of primary gaseous fuel compositions used in this study are given in Table 1 and Table 2, respectively.

The in-cylinder pressure history of combustion cycles was measured with KISTLER-6052C pressure transducer in conjunction with a 0.5° crank-angle encoder to identify the piston location. The rate of heat release (ROHR) was calculated using this equation [17]:

$$\frac{dQ}{d\theta} = \frac{\gamma}{\gamma-1} p \frac{dV}{d\theta} + \frac{1}{\gamma-1} V \frac{dp}{d\theta} \tag{1}$$

where θ is the crank angle (CA), p is the in-cylinder pressure at a given crank angle, V is the cylinder volume at that point and γ is the specific heat ratio. The ROHR represents the rate of energy release from the combustion process. The combustion transition from the first stage (slow flame propagation) to the second stage (end-gas autoignition) is identified from the ROHR. CO and NOx emissions were measured with a four-component analyzer Horiba PG-240, smoke was measured with a smoke meter Horiba MEXA-600s and HC emissions were measured with Horiba MEXA-1170HFID.

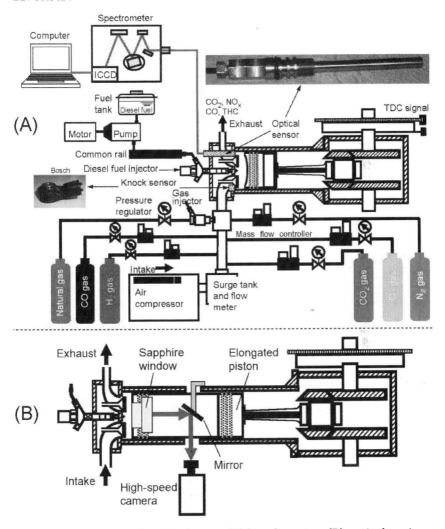

Figure 1. Experimental engine layout. (A) bench engine, (B) optical engine

An elongated cylinder liner and elongated piston were installed on the engine, Figure 1 (B), to visualize dual-fuel combustion events and to capture images of combustion in an optical engine. The engine mentioned above in Figure 1 (A) was modified to allow facilitating the visualization experiments.

Table 1. Experimental conditions

Engine type	4-stroke, single cylinder, water cooled
Bore x Stroke	96x108 mm
Swept volume	781.7 cm3
Compression ratio	16
Combustion system	Dual-fuel, direct injection
Combustion chamber	Shallow dish
Engine speed	1000 rpm
Intake pressure	101 kPa, 200 kPa
Injection system	Common-rail
Pilot fuel injection pressure	
- *Natural gas case*	40 MPa, 80 MPa, 120 MPa, 150 MPa
- *Syngas case*	80 MPa
Pilot fuel injection quantity	
- *Natural gas case*	2 mg/cycle, 3 mg/cycle
- *Syngas case*	3 mg/cycle
Nozzle hole x diameter	
- *Natural gas case*	3x0.08 mm, 3x0.10 mm, 4x0.10 mm
- *Syngas case*	4x0.10 mm
Equivalence ratio	
- *Natural gas case*	0.6
- *Syngas case*	Variable
EGR rate	
- *Natural gas case*	10%, 20%, 30%, 40%, 50%
- *Syngas case*	none
EGR composition	N_2-86%, O_2-10%, CO_2-4%

Table 2. Gas composition

Gas type	Composition						
	H₂ (%)	CO (%)	CH₄ (%)	CO₂ (%)	N₂ (%)	LHV (MJ/kg)	Source
Type 1	13.7	22.3	1.9	16.8	45.3	4.13	BMG
Type 2	20.0	22.3	1.9	16.8	39.0	4.99	BMG
Type 3	56.8	22.3	1.9	16.8	2.2	13.64	COG
Type 4	13.7	22.3	1.9	23.0	39.1	3.98	BMG
Type 5	13.7	22.3	1.9	34.0	28.1	3.74	BMG
Type 6	56.8	5.9	29.5	2.2	5.6	38.69	COG
Type 7	56.8	29.5	5.9	2.2	5.6	20.67	COG
Type 8	100.0	-	-	-	-	119.93	Hydrogen
	CH₄ (%)	C₂H₆ (%)	C₃H₈ (%)	n-C₄H₁₀ (%)		LHV (MJ/kg)	
Natural gas	88.0	6.0	4.0	2.0		49.20	

A sapphire window was installed on the top of the elongated piston. Temporal and spatial evolutions of visible flames were investigated by acquiring several images per cycle with MEMRECAM fx-K5, a high speed digital camera with the frame rate of 8000 fps in combination with a 45° mirror located inside the elongated piston. An engine shaft encoder and delay generator were used to implement camera-engine synchronization. The engine was first operated with motoring, then it was fired on only one cycle during which combustion images were captured.

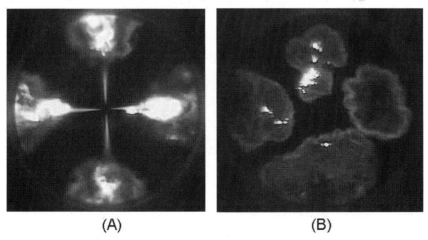

(A)   (B)

Figure 2. Dual fuel combustion of syngas with different amounts of injected diesel fuel (A) 10 mg and (B) 2 mg

Figure 2 shows the effect of pilot diesel fuel amount on combustion of syngas as a primary fuel. A distinct separation of diesel diffusion flame and syngas premixed flame is seen when larger amount of diesel fuel, 10 mg, is injected. The mixture burns only when the required oxygen is present. As a result, the flame speed is limited by the rate of diffusion. For syngas well-premixed mixture, the flame is not limited by the rate of diffusion and the mixture burning rate is much faster than that of diesel-air mixture. When only 2 mg of diesel fuel is injected the fuel evaporates and mixes with the oxidizer much faster providing distributed ignition centers in the cylinder for syngas-air mixture which is burnt as a premixed flame propagating towards the cylinder wall.

Figure 3 shows the dual-fuel combustion with natural gas (A) and syngas (B) as a primary fuel. The flame area growth for syngas is slower due to the presence of CO_2 and CO. Although syngas contains H_2, the effect of H_2 on flame propagation is not clearly seen due to simultaneous oxidation of H_2 and CO and the effect of CO_2 diluting the air-fuel mixture. The effect of CO oxidation in the presence of H_2O on flame propagation is a quite complex topic and is outside the scope of this chapter.

3. PREMIER MICRO-PILOT IGNITED DUAL-FUEL COMBUSTION

3.1. Concept of premier combustion

Before giving a description to PREMIER (PREmixed Mixture Ignition in the End-gas Region) combustion, it is necessary to explain the differences of phenomenological outline between conventional combustion and knocking combustion in the dual-fuel engine. Conventional combustion is a combustion process which is initiated by a timed pilot ignited fuel and in which the multiple flame fronts caused by multiple ignition centers of pilot fuel, moves completely across the combustion chamber in a uniform manner at a normal velocity. Knocking combustion is a combustion process in which some part or all of the charge may be consumed (autoignited) in the end-gas region at extremely high rates. Much evidence of end-gas mixture auto-ignition followed by knocking combustion can be obtained from high-speed laser shadowgraphs [18], high-speed Schlieren photography [19], chemiluminescent emission [20], and laser-induced fluorescence [21]. In addition, Stiebels *et al.* [22] and Pan and Sheppard [23] showed that multiple autoignition sites occur during knocking combustion. The combustion mode we have monitored we believe differs from knocking combustion in terms of the size, gradients, and spatial distribution of the exothermic centers in the end-gas. This combustion concept was given a name PREMIER combustion.

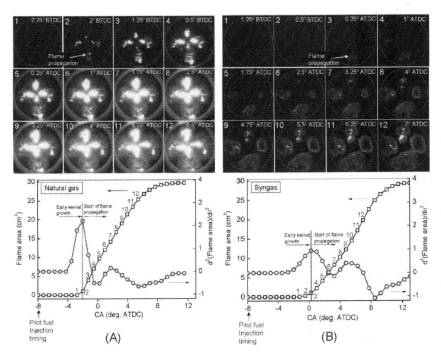

Figure 3. Dual-fuel combustion sequential images at P_{inj} = 40 MPa, P_{in} = 101 kPa, θ_{inj} = 8° BTDC, m_{df} = 2 mg/cycle, ϕ_t = 0.6. (A) natural gas, CH_4=88%, C_2H_6=6%, C_3H_8=4%, $n\text{-}C_4H_{10}$=2% (B) syngas, H_2=22.3%, CO=27.6%, CH_4=2.7%, CO_2=23.2%, N_2=24.2%

A conceptual outline of PREMIER combustion is presented in Figure 4. In the first stage of this combustion mode, the pilot diesel fuel is injected, evaporated, and auto-ignited prior to top dead center (TDC). The energy released by the diesel fuel auto-ignition initiates the gaseous flame development and outward propagation from the ignition centers toward the cylinder wall. Once the end-gas region is sufficiently heated and the temperature of the fuel mixture has reached the auto-ignition temperature of the gaseous fuel/air mixture after TDC, the second-stage combustion begins and is completed as the gas expands and cools, producing work. The second-stage heat release occurs over a chemical reaction timescale and is faster than heat release by turbulent flame propagation. Thus, the combustion transition from the first stage to the second stage takes place when the overall heat release rate changes from the slower first-stage flame rate to the faster second-stage rate, and that transition is here measured as the point where the second derivative of the heat release rate is maximized, as shown in Figure 4.

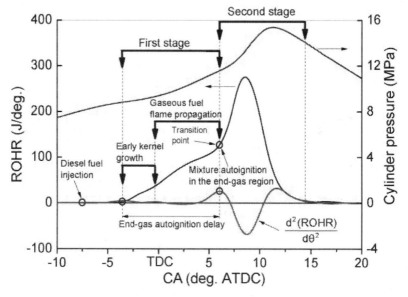

Figure 4. PREMIER combustion concept

PREMIER combustion in a dual-fuel engine is comparable to combining SI and CI combustion, which is being investigated by several researchers [24-26]. One disadvantage of these combustion strategies is that they are difficult to control under lean mixture conditions. The spark discharge is very short, and under light load and lean mixture conditions, the flame is too weak to propagate strongly and may be extinguished. Therefore, the combustion chamber must be specially designed to facilitate a stratified mixture charge around the spark plug electrodes. In a dual-fuel engine, on the other hand, combustion of the injected diesel fuel proceeds concurrently with that of the gaseous fuel mixture. This slow combustion of the diesel fuel helps to maintain the natural gas flame propagation and prevents the misfires that may occur under lean mixture conditions. The lean limit for the gaseous fuel/air mixture is of practical importance, as lean operation can result in both higher efficiency and reduced emissions. The major benefit of lean operation is the accompanying reduction in combustion temperature, which leads directly to a significant reduction in NOx emissions. The lean limit is the point where misfire becomes noticeable, and it is usually described in terms of the limiting equivalence ratio ϕlim that supports complete combustion of the mixture. For example, with the natural gas, if we operate slightly above the limiting equivalence ratio (for methane, $\phi_{lim} \approx 0.5$ [27]), the mixture reactivity becomes very sensitive to even very small variations in the air–fuel ratio. This high sensitivity is due to the presence of n-butane in the natural gas. It is known that small changes in the volume fraction of n-butane strongly affect the ignition properties of natural gas [28-31]. During hydrocarbon fuel oxidation, an H-atom is more easily abstracted from an n-butane molecule (with two secondary carbon atoms) than from other hydrocarbons such as methane, ethane, or propane [32]. As the equivalence ratio increases, the n-butane mass fraction in the natural gas/air mixture increases proportionally. During a fuel oxidation reaction, the in-cylinder gas temperature rises, and more of the radicals that initiate methane oxidation are created by increasing the ratio of n-butane. Similar results were documented by other researchers who investigated the auto-ignition and combustion of the natural gas in an HCCI engine [33]. They found that very small increases in the equivalence ratio of the methane/n-butane/air mixture produced significant changes in the profiles of the in-cylinder pressure traces.

If not extinguished during the early combustion stage, the pilot diesel flame may continue to a later stage, and pilot flame energy contributes to the stability of the combustion process [34]. The remaining unburned in-cylinder mixture from the first stage, located beyond the boundary of the flame front, is then subjected to a combination of heat and pressure for certain duration. As the flame front propagates away from the primary ignition points, end-gas compression raises the end-gas temperature and pressure. When the mixture is preheated throughout the combustion chamber volume and the end-gas mixture reaches the auto-ignition point, simultaneous auto-ignition occurs in several limited locations (known as exothermic centers), with a sharp increase in heat release. This part of the combustion appears as a rapid energy release in the second stage of the heat release curve shown in Figure 4.

A prime requirement for maintaining PREMIER combustion mode in a dual-fuel engine is that the mixture must not auto-ignite spontaneously during or following the rapid release of pilot energy. Failure to meet this requirement can lead to the onset of knock, which manifests itself in excessively sharp pressure increases and overheating of the walls, resulting in significant loss of efficiency with increased cyclic variations. When much smaller pilots are used, the energy release during the initial stages of ignition and the resulting turbulent flame propagation can (under certain conditions) lead to auto-ignition of the charge well away from the initial ignition centers, in the end-gas regions ahead of the propagating flames. This can occur in a manner that resembles the occurrence of knock in spark-ignition engines, but with controlled heat release. For the sake of convenience, the total energy release rate during PREMIER combustion can be divided into three sequential components. The first of these is due to the ignition of the pilot fuel. The second is due to the combustion of the gaseous fuel in the immediate vicinity of the ignition centers of the pilot, with consequent flame propagation. The third is due to auto-ignition in the end-gas region.

4. PREMIER COMBUSTION DETECTION (NATURAL GAS)

4.1. Cyclic variations and fft of in-cylinder pressure

Cycle-to-cycle variations of the in-cylinder pressure during conventional, PREMIER, and knocking combustion with natural gas are identified by overplotting 80 cycles of a measured pressure trace. From Figure 5, we observe that cycle-to-cycle pressure variations were present in all cases considered, and they varied according to the injection timing θ_{inj}. It should be noted that under normal combustion conditions, the magnitude of the peak pressure (which is directly related to the power output) depends on θ_{inj}, and P_{max} is higher for advanced θ_{inj}. Unfortunately, the advantage of this peak pressure increase is offset by the disadvantages of increased fluctuation and the occurrence of knock.

Thus, when slow combustion dominates, the P_{max} fluctuations are small, whereas during fast combustion, the P_{max} fluctuations are larger. With advanced fuel injection timing, the combustion was too fast, and the mixture auto-ignited spontaneously during or following the rapid pilot energy release. This led to the onset of knock. However, with a slight retardation of the injection timing, the energy release during the initial stages of ignition and the resulting turbulent flame propagation may induce auto-ignition of the charge in the end-gas regions ahead of the propagating flame, followed by PREMIER combustion. The stability of PREMIER combustion was confirmed by running an engine mentioned in Figure 1 (A) continuously for 30 minutes. As the engine operation reached a steady condition, P_{max} stabilized within a definite range between conventional and knocking combustion modes. It should be noticed that the present achieved stability of PREMIER combustion allows using this mode in stationary engines, such as engines used for power generation. In order to utilize PREMIER combustion in vehicles, further research is required to find optimum ways to precisely control the conditions inside the cylinder at various loads.

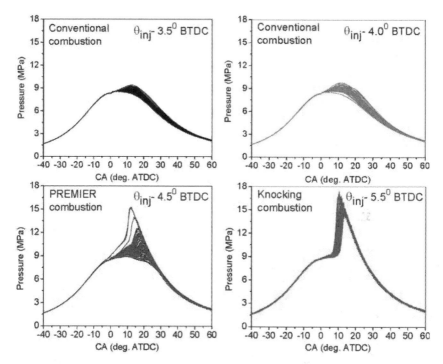

Figure 5. Cyclic maximum pressure versus its angle (Pmax, θ_{Pmax}) for different injection timings. P_{inj}=80 MPa, P_{in}=200 kPa, D_{hole}= 0.1 mm, N_{hole}=3, m_{df}=2 mg/cycle, ϕ_t=0.6

The transition from PREMIER combustion to knocking combustion was evaluated by fast Fourier transform (FFT) analysis of the in-cylinder pressure and the knock-sensor signal. The details of FFT analysis are given in [35]. Unlike during knocking in traditional spark-ignition engines, high frequency oscillations of in-cylinder pressure and knock-sensor signals were not observed during the PREMIER combustion mode.

The in-cylinder pressure and knock-sensor signals were filtered using an FFT band-pass filter with 4-20 kHz cutoff frequencies. The filtered pressure data was used to define the knock intensity. Figure 6shows that during PREMIER combustion mode, the measured cylinder pressure curve was smooth, and the maximum rate of pressure rise (which was only 0.36 MPa/CA deg.) occurred at 17° ATDC under the conditions P_{int}=200 kPa, m_{df}=2 mg/cycle, P_{inj}=80 MPa and θ_{inj}=4.5° BTDC. Pressure oscillations did not occur, and the pressure sensor was unable to detect a sawtooth pattern on the measured pressure trace. For the same conditions mentioned above, but at θ_{inj}=5° BTDC, the transition from PREMIER to knocking combustion occurred at the maximum rate of pressure increase (dP/dθ=0.55 MPa/CA deg.), with a knock intensity of K_{INT}=0.3 MPa. At $θ_{inj}$=5.5° BTDC, strong knocking combustion was detected with dP/dθ=1.87 MPa/CA deg. and K_{INT}=1.45 MPa. As the figure indicates, oscillations of the in-cylinder pressure and the knock sensor signal were not detected during PREMIER combustion mode. Weak oscillations were detected during the transition from PREMIER to knocking combustion, and stronger oscillations were detected at 6.52 kHz and 10.1 kHz during knocking combustion. The peak at the transition to knocking combustion was correspondingly lower than the peak that occurred during heavy knocking combustion.

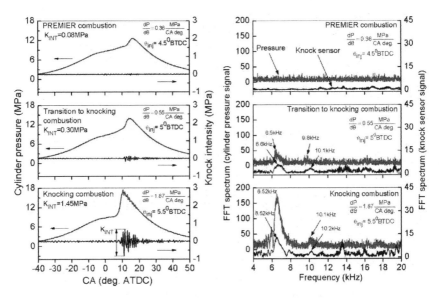

Figure 6. FFT analysis of in-cylinder pressure. P_{inj}=80 MPa, P_{in}=200 kPa, D_{hole}= 0.1 mm, N_{hole}=3, m_{df}=2 mg/cycle, $φ_t$=0.6

4.2. Spectroscopy Analysis

The configuration of our optical engine made it difficult to directly visualize auto-ignition in the end-gas region with an optical engine setup shown in Figure 1 (B). Since auto-ignition in the end-gas region usually occurs closer to the cylinder wall, the auto-ignition region was hidden from the camera view by the top of the elongated piston, where a sapphire window was installed. Thus, an optical sensor was inserted in the region based on the most probable occurrence of auto-ignition in the end-gas region, as shown in Figure 7. The small effect of measurement location on autoignition and flame development was confirmed based on several preliminary experiments. Besides, this measurement location was selected to keep the sensor away from the highly luminescent soot radiation. If the sensor is placed too close to the luminous emissions from diesel combustion, these emissions can supersede the OH* radical emissions, which are expected to occur during PREMIER combustion. Chemical luminescence emissions from the propagating flame and end-gas region auto-ignition were measured using a spectrometer equipped with intensified charge-coupled device (ICCD). The regions of spectroscopy measurements for conventional, PREMIER and knocking combustion, along the crank angle degrees, are shown in Figure 8. In-cylinder pressure and the rate of pressure rise change as injection timing is advanced for conventional, $\theta_{inj}=4º$ BTDC, PREMIER, $\theta_{inj}=8º$ BTDC and knocking, $\theta_{inj}=13º$ BTDC, combustion. Exposure time of the spectrometer for all three combustion regimes was set to 3º CA. The highlighted region on each graph shows the interval within which the measurements were taken.

Figure 9 shows the background light-subtracted ensemble-averaged spectra obtained between 200 and 800 nm as the propagating flame arrived in the vicinity of the optical sensor and auto-ignition occurred in the end-gas region. The background light was subtracted by the signal processing using a percentile filter. The waveforms of the emission spectroscopy were obtained for knocking cycles in the range of 0º-3º, 3º-6º, and 6º-9º ATDC, for PREMIER cycles in the range of 6º-9º, 9º-12º, and 12º-15º ATDC, and for conventional cycles in the range of 12º-15º, 15º-18º, and 18º-21º ATDC.

High OH* radical emission intensities were evident at wavelength 310 nm and very weak intensities at wavelength 286 nm when the flame front reached the optical sensor location region. OH* radical emission intensities at this wavelengths were stronger for knocking combustion cycles than for PREMIER combustion cycles, and for conventional combustion cycles these emission intensities were not seen. A similar trend was observed by Itoh *et al.* [36] and Hashimoto *et al.* [37]. These authors reported that under non-knocking operation, OH* radicals exhibit comparatively weak emission intensities. However, the OH* radical emission intensity gradually increased for autoignition and knocking cycles in comparison with a conventional cycle.

4. PREMIER combustion detection (Natural gas)

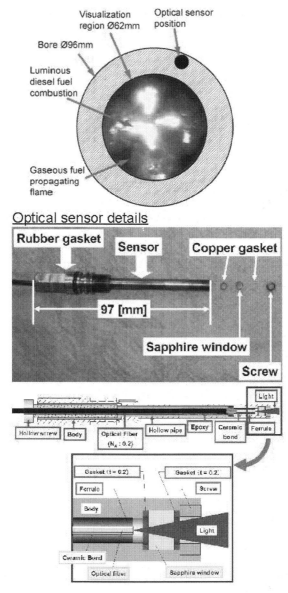

Figure 7. Optical sensor location and design

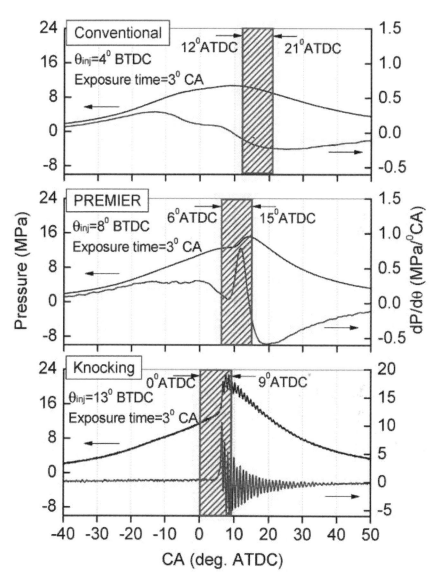

Figure 8. In-cylinder pressure and the rate of pressure rise. P_{inj}=40 MPa, P_{in}=200 kPa, D_{hole}=0.1mm, N_{hole}=3, m_{df}=2 mg/cycle, ϕ_t=0.6

Figure 9 shows that for the knocking cycle, the OH* peak at 310 nm increased sharply at 6º ATDC and then decreased at 9º ATDC. 6º ATDC corresponds to the timing of the maximum rate of pressure rise, and 9º ATDC corresponds to the timing when the rate of pressure rise is gradually decreased. Before natural gas propagating flame (1st stage combustion) is stabilized, the rapid autoignition occurs in multiple spots. The end-gas autoignition kernel growth rate is much faster than that of flame propagation.

The drastic increase in OH* emissions is related to the sharp pressure and temperature increases in the end-gas region. It has been reported that the concentration of OH* radicals shows a strong temperature dependence in the thermal ignition region [38]. The thermal ignition is the hydrogen-oxygen system reactions, and the CO conversion into CO_2 with the assistance of OH* takes place simultaneously with the hydrogen-oxygen system reactions. It has been observed that the OH* emission intensity at auto-ignition shows the same tendency as the correlation between the occurrence of auto-ignition and the knocking intensity [39]. On the other hand, the OH* peaks of PREMIER combustion at 12° and 15°ATDC have the same magnitude. 12° ATDC corresponds to the timing of the maximum rate of pressure rise, and 15° ATDC corresponds to the timing after the peak of the rate of pressure rise. The reason for the same magnitude in OH* is that the maximum rate of pressure rise is slower than that of during knocking case. The end-gas autoignition kernel growth rate is comparable with that of flame propagation. The mixture in the end-gas region reacts steadily with the steady OH*emission.

The emissions at the wavelengths above 700 nm are due to diesel luminous flame. The previous shock tube measurements showed that continuum emission at wavelengths above 650 nm was due to either young soot particles or large hydrocarbon molecules [40]. Zhao and Ladommatos [41] showed that the maximum emissive power of soot particles, estimated as a blackbody, occurs in the range of about 680–1100 nm. Vattulainen et al. [42] confirmed that the light emitted by a diesel flame is dominated by soot incandescence, and the emission range was determined to be about 650–800 nm. The emissions at the wavelengths above 700 nm shown in Figure 9 correspond to 1-5% of the total wavelength emission band from a blackbody at the estimated average peak burned-gas temperature of 2310 K under knocking conditions, which correspond to 1.823×10^5-4.935×10^5 W/m² µm of spectral emissive power, with the maximum spectral emissive power at that temperature that equals to 8.459×10^5 W/m² µm [43].

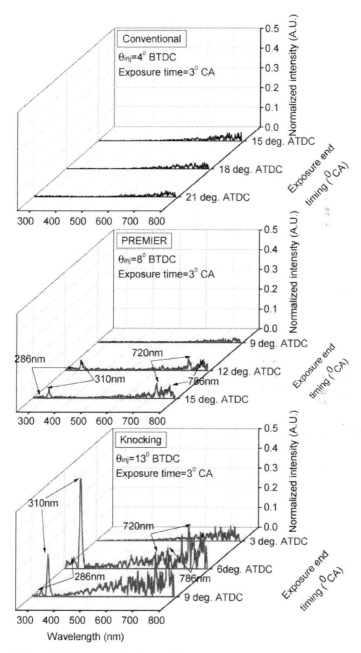

Figure 9. Detailed spectra in the end-gas region. P_{inj}=40 MPa, P_{in}=200 kPa, D_{hole}=0.1 mm, N_{hole}=3, m_{df}=2 mg/cycle, ϕ_t=0.6

5. PREMIER COMBUSTION CHARACTERISTICS

5.1. Natural Gas Combustion Characteristics

To maintain PREMIER combustion in a dual-fuel natural gas engine, the effects of several operating parameters must be identified. Our experimental results show that the major parameters that may significantly influence the energy release pattern during dual-fuel PREMIER combustion are pilot diesel fuel injection timing and the EGR rate, which can affect the total equivalence ratio based on oxygen content. Other parameters such as pilot fuel injection pressure, injected pilot fuel amount, nozzle hole diameter, and hole number have minor effects on PREMIER combustion.

5.1.1. Effect of injection timing

As shown in Figure 10, PREMIER combustion can be maintained within a wide range of pilot fuel injection pressures. However, it can be maintained within only a very narrow range of fuel injection timings. At a fixed total equivalence ratio, advancing the injection timing resulted in the earlier occurrence of combustion during the cycle, increasing the peak cylinder pressure during first-stage combustion. With the burned gas of first-stage combustion, the in-cylinder pressure and temperature continued to rise after TDC, as shown in Figure 10. Although the piston began to move downward after TDC, and the volume thus expanded, the heat release from first-stage combustion induced local temperature and pressure increases during second-stage combustion. Higher peak cylinder pressures resulted in higher peak charge temperatures. Retarding the injection timing decreased the peak cylinder pressure during first-stage combustion, as more of the fuel burned after TDC.

Advancing the injection timing resulted in better diesel fuel evaporation and mixing with the in-cylinder gas. Therefore, diesel fuel auto-ignition occurred more quickly and with more complete diesel fuel combustion and natural gas flame propagation during the first stage, resulting in rapid combustion and high heat release rate during the second stage due to the rapid heating of the end-gas region mixture.

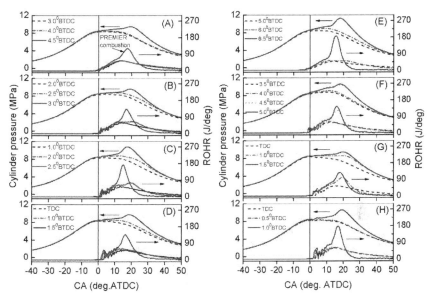

Figure 10. Effect of pilot fuel injection timing on cylinder pressure and the rate of heat release. Experimental conditions are given in Table 3.

Table 3. Experimental conditions for Figure 10.

	Injection pressure P_{inj} (MPa)	Intake pressure P_{in} (kPa)	Nozzle hole diameter D_{hole} (mm)	Nozzle hole number N_{hole}	Injected pilot fuel amount m_{df} (mg/cycle)
A	40	200	0.1	3	3
B	80	200	0.1	3	3
C	120	200	0.1	3	3
D	150	200	0.1	3	3
E	40	200	0.1	3	2
F	80	200	0.1	3	2
G	150	200	0.08	3	3
H	150	200	0.1	4	3

5.1.2. Mass fraction burned in the second stage of rohr

Figure 11 shows a relation between the rate of maximum pressure rise and the second stage autoignition delay. The second stage autoignition delay time was estimated as a time between two peaks of the second derivative of the ROHR, as shown in Figure 4. Figure 11 also shows that for the range of

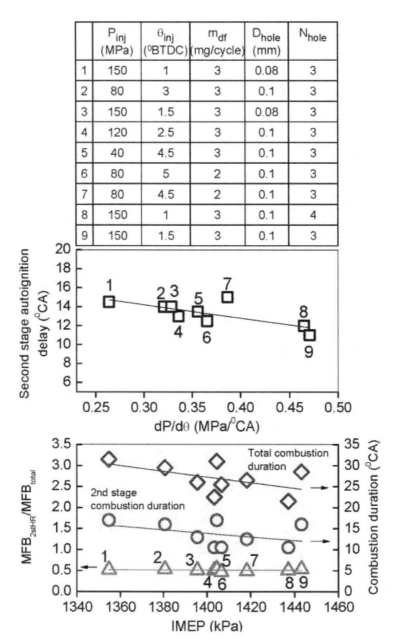

Figure 11. Second stage autoignition delay and mass fraction burned during the second stage of combustion of natural gas

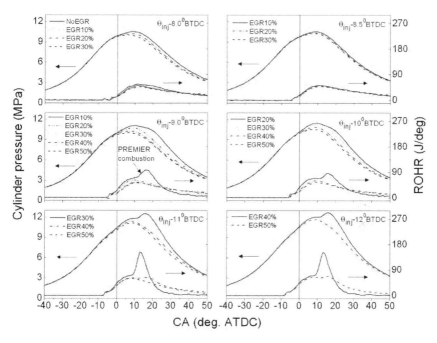

Figure 12. Effect of EGR on cylinder pressure and the rate of heat release. P_{inj}=40 MPa, P_{in}=200 kPa, D_{hole}=0.1 mm, N_{hole}=3, m_{df}=2 mg/cycle

experimental conditions the mass fraction burned during the second stage of the ROHR is remained nearly the same, although IMEP increases. This increase in IMEP is due to faster and more intense combustion with the shorter duration in both first and second stage of the ROHR. Reduced heat loss during shorter combustion duration ensures higher in-cylinder pressure and temperature, and therefore, higher IMEP. This trend suggests that the second stage combustion is influenced by the first stage. Although the mass fraction burned in the second stage is the same, the combination of several operational parameters such as pilot fuel injection pressure, injection timing, the amount of injected pilot fuel, gaseous fuel equivalence ratio and nozzle characteristics may affect, to a certain degree, the progress of PREMIER combustion.

5.1.3. Effect of egr

PREMIER combustion becomes clearly recognizable if the EGR rate remains below a certain level. When the EGR rate surpasses this level, the unburned mixture temperature decreases, retarding the combustion of the natural gas and affecting the reactivity of the mixture to auto-ignite in the end-gas region. As Figure 12 shows, at 200 kPa of intake pressure and moderate EGR rates, the first-stage combustion rate increased, and second-stage heat release was able to occur. A similar trend was also observed at 101 kPa of intake pressure, although not shown here. These results suggest that the use of EGR may not be advantageous for achieving PREMIER combustion. However, it should be noted that for engine operation close to the knock-limit conditions, the high combustion rate of natural gas may be markedly decreased by using a certain limited EGR rate, and maintaining PREMIER combustion mode as the knocking effect is suppressed.

5.2. Natural Gas Exhaust Emission Characteristics

Figure 13 shows the engine performance characteristics and exhaust gas emissions. The encircled data on (A) and (B) of the figure correspond to PREMIER combustion. During PREMIER combustion, a considerable increase in the indicated mean effective pressure and thermal efficiency was observed. This was due to the sharp increase of the second heat-release peak within a shorter crank-angle time, as seen in Figures 10 and 12. Moreover, this implies that the total combustion time for PREMIER combustion mode was shorter than the total time required for conventional combustion. Although HC and CO emissions were greatly decreased, NOx emissions increased considerably compared with conventional combustion. This increase in NOx emissions is expected, as in-cylinder temperature increase hastens the oxidation reactions of in-cylinder nitrogen and oxygen.

5.3. Syngas Combustion Characteristics

Figure 14 indicates the relationship between the rate of maximum pressure rise and the second stage autoignition delay. The second stage autoignition delay time was estimated as time between two peaks of the second derivative of the ROHR as shown in Figure 4. It was found that for all types of gases investigated in this chapter the autoignition delay of the second stage decreases with the increase of the rate of maximum pressure rise.

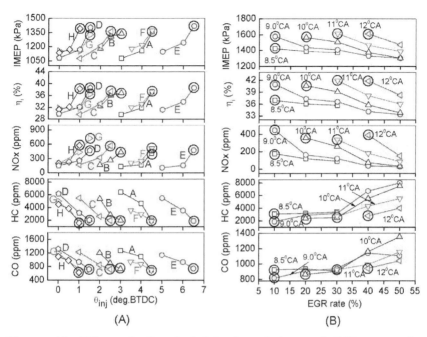

Figure 13. Effect of pilot fuel injection timing and EGR on engine performance and emissions. (A) conditions correspond to those of Figure 10, (B) conditions correspond to those of Figure 12.

Figure 15 shows the in-cylinder pressure and ROHR for Type 1 (A) and Type 2 (B) of syngas at different equivalence ratios and various injection timings. The results show that the maximum pressure and heat release rate reached higher values for the gas with higher H_2 content. As the injection timing was gradually advanced, PREMIER combustion with second-stage heat release occurred. PREMIER combustion was observed at various equivalence ratios for certain injection timings. For instance, for Type 1 gas, two-stage heat release appeared at ϕ = 0.4 starting from 23º BTDC, at ϕ = 0.52 from 15º BTDC, at ϕ = 0.68 from 9º BTDC, and at ϕ = 0.85 from 7º BTDC. The same trend was observed for Type 2 gas, but the maximum heat release rate was higher than that of Type 1, due to the higher H_2content. The maximum cylinder pressure for Type 2 decreased at an equivalence ratio of 0.83, since the injection timing in that case needed to be retarded to around TDC (and even to the expansion stroke) to avoid knock. At the same time, for Type 1 gas at an equivalence ratio of 0.85 and injection timings before TDC, PREMIER combustion was clearly observed without any knock. These results seem to suggest that the increased H_2 content of Type 2 gas affected the ignitability and corresponding progress of combustion that leads to engine knock.

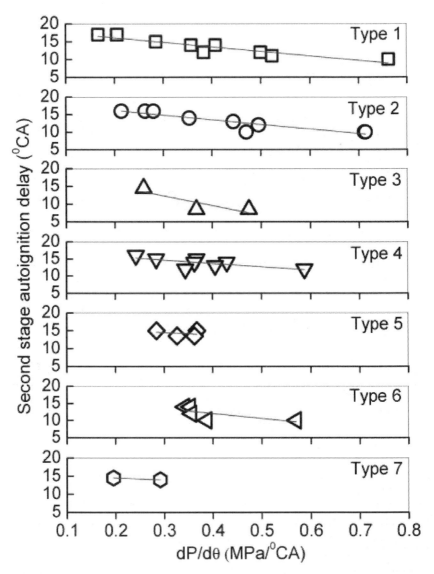

Figure 14. Second stage autoignition delay. P_{inj}=80 MPa, P_{in}=200 kPa, m_{df}=3 mg/cycle

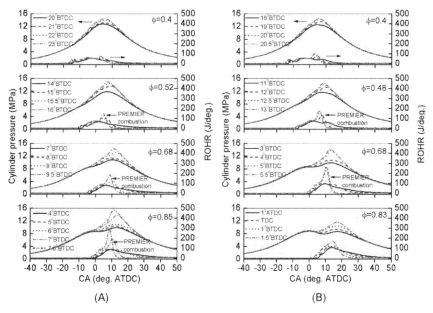

Figure 15. Comparison of in-cylinder pressure and ROHR. Syngas, (A) Type 1, (B) Type 2

5.3.1. Effect Of H_2 Content On Premier Combustion

An increase in hydrogen content of syngas results in an increase of ignitability and a corresponding reduction in ignition delay of the first stage. To investigate the effect of mass fraction burned in the second stage and the effect of hydrogen content on the rate of maximum pressure rise, the ratio MFB_{2stHR}/MFB_{total} was evaluated. MFB_{2stHR} is the integral of the mass fraction burned during the second stage, computed from the transition point where the peak of $d^2(ROHR)/d\theta^2$ is maximized to the 80% MFB. MFB_{total} is the integral of the total mass fraction burned, computed from the first peak of $d^2(ROHR)/d\theta^2$ to the 80% MFB. Figure 16 shows that as the mass fraction burned increases the rate of maximum pressure rise also increases. The same trend was monitored at various equivalence ratios.

Figure 17 shows the effect of H_2 content on the mean combustion temperature, IMEP, indicated thermal efficiency and NOx for conventional and PREMIER combustion at the same input energy Q_{in}=2300 J/cycle and injection timing at the minimum advance for the best torque (MBT). As this figure shows, the increase in H_2 amount affects the engine combustion characteristics. For PREMIER combustion, the mean combustion temperature and consequently the NOx significantly increase when compared with those of conventional combustion. IMEP and indicated thermal efficiency increase about 10%. Therefore, in dual-fuel combustion of low-energy density syngas, PREMIER combustion is an important combustion mode that tends to increase the engine efficiency.

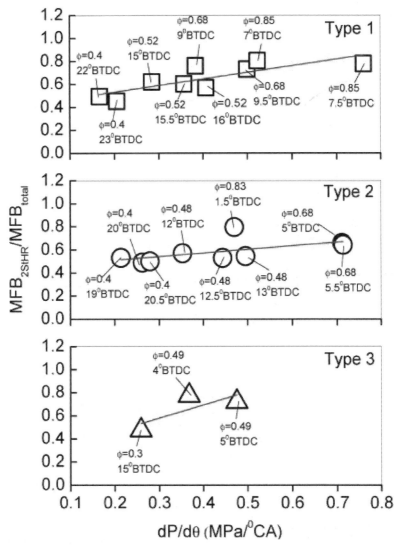

Figure 16. Fuel mass fraction burned with the change of H_2 content

5.3.2. Effect Of CO₂ Content On Premier Combustion

An increase in CO_2 content in syngas results in a dilution of the mixture with the corresponding reduction in the rate of fuel oxidation reactions and consequent combustion. To investigate the effect of mass fraction burned in the second stage and the effect of CO_2 content on the rate of maximum pressure rise, the same procedure was applied as explained in the previous section.

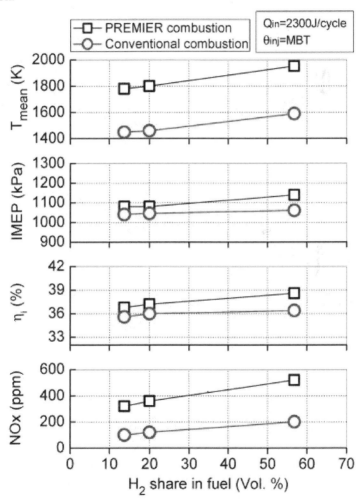

Figure 17. Effect of H_2 content on engine performance characteristics at θ_{inj}=MBT

Figure 18 shows that for Type 1 and Type 4 as the mass fraction burned during the second stage increases the rate of maximum pressure rise also increases. However, for Type 5 with 34% of CO_2 content in syngas, the rate of maximum pressure rise decreases with the increase of the mass of fuel burned in the second stage. This can be explained by the fact that although the total mass of syngas burned during the second stage increases, the CO_2 fraction in the gas also proportionally increases. Eventually, the certain threshold can be reached when the effect of CO_2 mass fraction in the gas on combustion overweighs the effect of H_2.

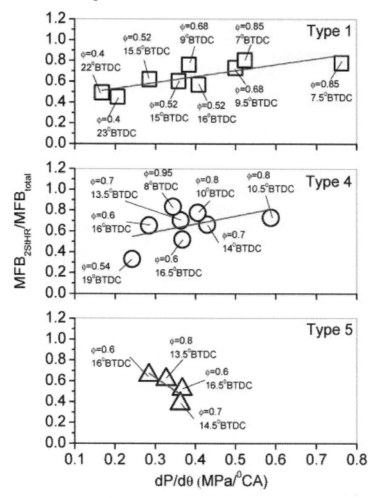

Figure 18. Fuel mass fraction burned with the change of CO_2 content

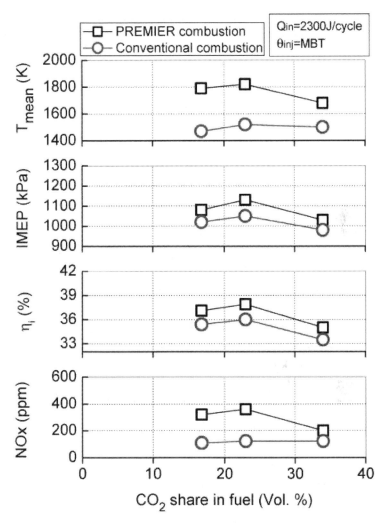

Figure 19. Effect of CO_2 content on engine performance characteristics at θ_{inj}=MBT

Figure 19 shows the effect of CO_2 content on the mean combustion temperature, IMEP, indicated thermal efficiency and NOx for conventional and PREMIER combustion at the same input energy Q_{in}=2300 J/cycle and injection timing at MBT. In this figure the trend mentioned earlier in Figure 18 is clearly observed.

The mean combustion temperature, IMEP, the indicated thermal efficiency and the NOx shows the increase when CO_2 content in the gas increases from

16.8% to 23%. However, as CO_2 concentration reaches 34%, above mentioned combustion characteristics decrease. This trend was observed for both conventional and PREMIER combustion. Therefore, in order to achieve high combustion efficiencies in dual-fuel engines fuelled with low-energy density syngas, CO_2 fraction in syngas needs to be controlled.

5.4. Syngas Exhaust Emission Characteristics

Increasing hydrogen content in syngas has a substantial impact on engine performance and pollutants formation. Engine performance characteristics and concentrations of pollutants CO, HC and NOx are shown in Figures 20 and 21. It should be noticed that the smoke level was negligibly low. Figure 3 (B) shows highly luminous diesel fuel droplets burn exposed to the high temperature syngas flame. This concurrent micro-pilot diesel fuel and syngas combustion contributes to the faster oxidation and burn out of soot by the end of the combustion process.

Figures 20 and 21 show the experimental operation region for IMEP, CO, HC and NOx. The comparison is given for H_2 effect between Type 1 and Type 2, for CO_2 effect between Type 1 and Type 5, and for gas - type effect between Type 6 (Coke-oven gas) and Type 8 (Pure hydrogen). The dash-dotted line is the boundary separated the conventional combustion from the PREMIER combustion at corresponding equivalence ratios and injection timings.

Figure 20 shows that higher IMEP levels appear at higher equivalence ratios and advanced injection timings. PREMIER combustion mode may pass through different IMEP levels, depending on the equivalence ratio and injection timing. As H_2 fraction increases in the syngas (from Type 1 to Type 2), PREMIER combustion region expands at lower equivalence ratios and shrinks at higher equivalence ratios. On the contrary, as CO_2 fraction increases in the syngas, from Type 1 (16.8% CO_2) to Type 5 (34% CO_2), the same maximum level of IMEP, as for Type 1 and 2, can be achieved with only higher equivalence ratios. Type 1 and Type 5 show that as CO_2 content increases IMEP slightly decreases. With the increase of CO_2 fraction in the syngas the PREMIER combustion region threshold is shifted towards the boundary of operation domain. This implies that the operational region of PREMIER combustion mode is reduced.

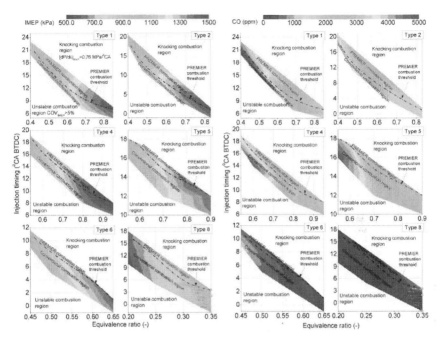

Figure 20. Experimental mapping of IMEP and CO distribution

Figure 20 (Type 1 and Type 2) and Figure 21 (Type 1 and Type 2) show that with the increase of H_2 concentration in the gas content, CO and HC emissions are reduced. The cause of these reductions is most likely the enhanced oxidation occurring because of improved combustion and higher concentrations of reactive radicals since the carbon content in the gas is the same for both types of fuel. The NOx emissions, as in Figure 21, are higher for Type 2 than those for Type 1 due to higher H_2 content. This is in part a direct result of hydrogen's higher flame temperature effect on NO formation chemistry. NOx emissions are consistently reduced by lowering equivalence ratio for both types. For Type 5, as CO_2 content in the gas increases to 34%, the NOx reduction trend is obvious.

The opposite trend is observed for Type 6, as shown in Figure 20. This type of gas has very high content of H_2-56.8%, and very low content of CO_2-2.2%. For the limited range of equivalence ratios and injection timings, the PREMIER combustion region is very narrow, and at even higher equivalence ratios, the stable combustion will easily turn to knocking. These results show that in order to achieve PREMIER combustion with the best case scenario in terms of efficiency and emissions the equivalence ratio and injection timing should be maintained within certain range, depending on H_2 and CO_2 fractions in syngas. Figure 21 shows that for Type 6 the NOx emissions can be compared with those of Type 2. Even with the retarding of the injection timing closer to TDC, NOx emissions level is still comparably high. On the other hand, CO emissions are significantly reduced.

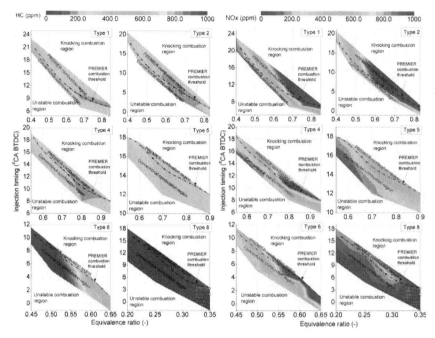

Figure 21. Experimental mapping of HC and NOx distribution

In addition, the performance and emissions of pilot-ignited dual-fuel engine operated on 100% H_2 as a primary fuel was investigated, shown as Type 8. The data at equivalence ratios of 0.2, 0.25 and 0.3 are related to the engine operation without dilution. Further increase in equivalence ratio above 0.3 caused knocking combustion. To further increase the energy supply from hydrogen the equivalence ratio of 0.35 was used with the 40%, 50% and 60% of N_2 dilution. Previous research suggests that NOx emissions in hydrogen-fueled engines can be reduced by using EGR or using the nitrogen dilution [44, 45]. These studies reported that NOx emissions were reduced by the reduction in peak combustion temperature due to the presence of diluent gas. Mathur *et al.* [45] showed that with larger concentrations of nitrogen as diluent, a greater amount of combustion heat can be absorbed, which results in reducing peak flame temperature and NOx. IMEP increases with the equivalence ratio. For the equivalence ratios of 0.2 and 0.25, the minimum advanced injection timing for the maximum IMEP was obtained at 17° BTDC and 12° BTDC, respectively. At these injection timings PREMIER combustion occurs. For the case with N_2 dilution, injection timing has greater effect on IMEP than the dilution rate. IMEP gradually increased as the injection timing was advanced along with the increased amount of diluent from 40% to 60%. Due to low equivalence ratios, very low level of NOx was detected without N_2 dilution. At the conditions with N_2 dilution the NOx showed further decrease. For the conditions without dilution, CO and HC emissions were significantly reduced to the level of about 6 ppm and 20 ppm, respectively. However, for the conditions with N_2 dilution CO varied from min.-15 ppm to max.-891 ppm and HC varied from min.-20 ppm to max.-125 ppm.

6. CONCLUSION

The new PREMIER combustion mode in a dual-fuel engine fuelled with natural gas, syngas and hydrogen was investigated via engine experiments. The following conclusions can be drawn from this research:

- PREMIER combustion combines two main stages of heat release, the first is gaseous fuel flame propagation and the second is end-gas mixture auto-ignition. The second stage can be mainly controlled by the pilot fuel injection timing, gaseous fuel equivalence ratio, and EGR rate. The delay time for mixture autoignition in the end-gas region is defined as the time from early kernel development to the transition point where slower combustion rate (flame propagation) is changed to faster combustion rate (autoignition). It was found that the rate of maximum pressure rise increases as the second stage ignition delay decreases. PREMIER combustion was observed for natural gas and all syngas types investigated in this paper. This type of combustion can enhance the engine performance and increase the efficiency.

- In both PREMIER and knocking combustion with natural gas, moderate and high intensities of OH* radicals were detected, respectively, at wavelengths of 286 and 310 nm when the flame front reached the optical sensor location region. In knocking combustion, the OH* radical emission intensities were stronger than those in PREMIER combustion, and in conventional combustion, these emission intensities could barely be detected. PREMIER combustion differs from knocking combustion in terms of size, gradients, and spatial distribution of exothermic centers in the end-gas region. High-frequency oscillation of the in-cylinder pressure did not occur during the PREMIER combustion mode with natural gas.

- An increase in the fuel mass fraction burned in the second stage of heat release affects the rate of maximum pressure rise. When hydrogen content in syngas is increased the same rate of maximum pressure rise can be achieved with lower amount of fuel mass fraction burned during the second stage, meaning that the increased amount of hydrogen in syngas induces an increase in the mean combustion temperature, IMEP and efficiency, but also a significant increase in NOx emissions. The results also show that when CO_2 content in the gas reaches 34%, the rate of maximum pressure rise, as well as, the mean combustion temperature, IMEP, thermal efficiency and NOx decrease despite the increase in fuel mass fraction burned during the second stage of heat release rate.

- For pure hydrogen at equivalence ratios of 0.2, 0.25 and 0.3 without dilution, very low CO and HC emissions were detected. The further increase in equivalence ratio above 0.3 led to knocking combustion. At the equivalence ratio of 0.35 with N_2 dilution, NOx level significantly decreased but CO level increased.

REFERENCES

1. C. S Weaver, Natural gas vehicles- a review of the state of the art, SAE Paper, 892133.
2. R. J Nichols, The challenges of change in the auto industry: Why alternative fuels? J Eng Gas Turb Power 199411672732
3. T Lieuwen, V Yang, R Yetter, Synthesis gas combustion: Fundamentals and applications. Taylor & Francis Group 2010
4. N. Z Shilling, D. T Lee, IGCC-Clean power generation alternative for solid fuels: GE Power Systems. PowerGenAsia 2003
5. Bade Shrestha SOKarim GA. Hydrogen as an additive to methane for spark ignition engine applications. Int J Hydrogen Energy 19992457786
6. R. F Stebar, F. B Parks, Emission control with lean operation using hydrogen-supplimented fuel. SAE Paper 740187.
7. L Jingding, G Linsong, D Tianshen, Formation and restraint of toxic emissions in hydrogen-gasoline mixture fueled engines. Int J Hydrogen Energy 19982397175

8. M Pushp, S Mande, development of 100% producer gas engine and field testing with pid governor mechanism for variable load operation. SAE Paper 2008

9. Y Yamasaki, G Tomatsu, Y Nagata, S Kaneko, Development of a small size gas engine system with biomass gas (combustion characteristics of the wood chip pyrolysis gas), SAE Paper 2007

10. Y Ando, K Yoshikawa, M Beck, H Endo, Research and development of a low-BTU gas-driven engine for waste gasification and power generation. Energy 200530220618

11. G. P Mctaggart-cowan, H. L Jones, S. N Rogak, W. K Bushe, P. G Hill, S. R Munshi, The effect of high-pressure injection on a compression-ignition, direct injection of natural gas engine. J Eng Gas Turb Power 200712957988

12. W Su, Z Lin, A study on the determination of the amount of pilot injection and rich and lean boundaries of the pre-mixed CNG/Air mixture for a CNG/Diesel dual-fuel engine. SAE paper, 2003

13. E Tomita, N Fukatani, N Kawahara, K Maruyama, T Komoda, Combustion characteristics and performance of supercharged pyrolysis gas engine with micro-pilot ignition. CIMAC congress 2007Paper 178

14. Z Liu, G. A Karim, Simulation of combustion processes in gas-fuelled diesel engine. J Power Energy 199721115969

15. F Ma, Y Wang, S Ding, L Jiang, Twenty percent hydrogen-enriched natural gas transient performance research. Int J Hydrogen Energy 200934642331

16. C. J Sung, C. K Law, Fundamental combustion properties of H2/CO mixtures: Ignition and flame propagation at elevated temperatures. Combust Sci.Tech. 200818010971116

17. J. B Heywood, Internal combustion engine fundamentals. New-York: McGraw-Hill 1988

18. Y Nakagawa, Y Takagi, T Itoh, and T Iijima, Laser shadowgraphic analysis of knocking in SI engine. SAE Paper, 845001.

19. J Pan, Sheppard CGW, Tindall A, Berzins M, Pennington SV and Ware JM. End-gas inhomogeneity, autoignition and knock. SAE Paper, 982616.

20. N Kawahara, E Tomita, and Y Sakata, Auto-ignited kernels during knocking combustion in a spark-ignition engine. Proc. Combust. Inst. 20073129993006

21. B Bauerle, F Hoffman, F Behrendt, and J Warnatz, Detection of hot spots in the end gas of an internal combustion engine using two-dimensional LIF of formaldehyde. In: 25th Symposium (Int.) on Combustion 199425135141

22. B Stiebels, M Schreiber, and Sadat Sakak A. Development of a new measurement technique for the investigation of end-gas autoignition and engine knock. SAE Paper, 960827.

23. J Pan, and Sheppard CGW. A theoretical and experimental study of the modes of end gas autoignition leading to knock in SI engines. SAE Paper, 942060.

24. J Li, H Zhao, N Ladommatos, Research and development of controlled auto-ignition (CAI) combustion in a 4stroke multi cylinder gasoline engine. SAE Paper, 2001

25. H Santoso, J Matthews, W. K Cheng, Managing SI/HCCI dual-mode engine operation. SAE Paper, 2005

26. H Persson, A Hultqvist, B Johansson, A Remon, Investigation of the early flame development in spark assisted HCCI combustion using high-speed chemiluminescence imaging. SAE Paper, 2007

27. F Ma, J Wang, Wang Yu, Wang Y, Li Y, Liu H, Ding S. Influence of different volume percent hydrogen/natural gas mixtures on idle performance of a CNG engine. Energy & Fuels 200822188087

28. Higgin RMRWilliams A. A shock-tube investigation of the ignition of lean methane and n-butane mixtures with oxygen. In: Symposium (Int.) on Combustion 19691257990

29. R Zellner, K. J Niemitz, J Warnatz, Gardiner Jr WC, Eubank CS, Simmie JM. Hydrocarbon induced acceleration of methane-air ignition. Prog Astronaut Aeronaut 19818825272

30. C. S Eubank, M. J Rabinovitz, Gardiner Jr WC, Zellner RE. Shock-initiated ignition of natural gas-air mixtures. In: 18th Symposium (Int.) on Combustion 198118176774

31. R. W Crossley, E. A Dorko, K Scheller, A Burcat, The effect of higher alkanes on the ignition of methane-oxygen-argon mixtures in shock waves. Comb. Flame 19721937378

32. C. K Westbrook, W. J Pitz, W. R Leppard, The autoignition chemistry of paraffinic fuels and pro-knock and anti-knock additives: a detailed chemical kinetic study. SAE Paper, 912314.

33. D Jun, K Ishii, N Iida, Autoignition and combustion of natural gas in a 4-stroke HCCI engine. JSME International J. 2003466067

34. H Saito, T Sakurai, T Sakonji, T Hirashima, K Kanno, Study on lean-burn gas engine using pilot oil as the ignition source. SAE Paper, 2001

35. U Azimov, E Tomita, N Kawahara, Y Harada, PREMIER (Premixed Mixture Ignition in the End-gas Region) combustion in a natural gas dual-fuel engine: Operating range and exhaust emissions. Int J Engine Research 201112484497

36. T Itoh, T Nakada, and Y Takagi, Emission characteristics of OH and C2 radicals under engine knocking. JSME International J. 199538230237

37. S Hashimoto, Y Amino, K Yoshida, H Shoji, and A Saima, Analysis of OH radical emission intensity during autoignition in a 2-stroke SI engine. In: Proceedings of the 4th COMODIA 19981998405410

38. A. G Gaydon, The spectroscopy of flames, 2nd edition, 1974Chapman and Hall Ltd, London).

39. T Itoh, T Takagi, and T Iijima, Characteristics of mixture fraction burned with autoignition and knocking intensity in a spark ignition engine. Trans. Japan Soc. Mech. Eng. 1985

40. C. M Coats, and A Williams, Investigation of the ignition and combustion of n-Heptane-oxygen mixtures. Proc. Combust. Institute 197917611621

41. H Zhao, and N Ladommatos, Optical diagnostics for soot and temperature measurement in diesel engines. Prog. Energy Combust. Sci. 199824221255

42. J Vattulainen, Experimental determination of spontaneous diesel flame emission spectra in a large diesel engine operated with different diesel fuel qualities. SAE Paper, 981380.

43. F. P Incropera, and D. P Dewitt, Fundamentals of heat and mass transfer, 4th edition, 1996John Wiley & Sons).

44. P. K Bose, D Maji, An experimental investigation on engine performance and emissions of a single cylinder diesel engine using hydrogen as induced fuel and diesel as injected fuel with exhaust gas recirculation. Int J Hydrogen Energy 200934484754

45. H. B Mathur, L. M Das, T. N Patro, Hydrogen-fuelled diesel engine: Performance improvement through charge dilution techniques. Int J Hydrogen Energy 19931842131

Chapter 9

Optimization of Diesel Engine with Dual-Loop EGR by Using DOE Method

Jungsoo Park[1] and Kyo Seung Lee[2]

[1] *The Graduate School, Department of Mechanical Engineering, Yonsei University, Sinchon-dong, Seodaemun-gu, Seoul, Korea*

[2] *Department of Automotive Engineering, Gyonggi College of Science and Technology, Jeongwang-dong, Siheung-si, Gyonggi-do, Korea*

1. INTRODUCTION

The diesel engine has advantages in terms of fuel consumption, combustion efficiency and durability. It also emits lower carbon dioxide (CO_2), carbon monoxide(CO) and hydrocarbons(HC). However, diesel engines are the major source of NOx and particulate matter emissions in urban areas. As the environmental concern increases, a reduction of NOx emission is one of the most important tasks for the automotive industry. In addition, future emission regulations require a significant reduction in both NOx and particulate matter by using EGR and aftertreatment systems (Johnson, 2011).

Exhaust gas recirculation (EGR) is an emission control technology allowing significant NOx emission reductions from light- and heavy-duty diesel engines. The key effects of EGR are lowering the flame temperature and the oxygen concentration of the working fluid in the combustion chamber (Zheng et al., 2004).

There are conventional types of EGR, high pressure loop and low pressure loop EGR. Table 1 shows advantages and drawbacks of each type of EGR.

1. Introduction

Table 1. Advantages and different types of EGR

	Advantages	Drawbacks
HPL EGR	• Lower HC and CO emissions • Fast response time	• Cooler fouling • Unstable cylinder-by-cylinder EGR distribution
LPL EGR	• High cooled EGR • Clean EGR (no fouling) • Stable cylinder-by-cylinder EGR distribution • Better Φ/EGR rate	• Corrosion of compressor wheel due to condensation water • Slow response time • HC/CO increase

Because a HPL EGR has fast response, especially at lower speed and load, it is only applicable when the turbine upstream pressure is sufficiently higher than the boost pressure. For the LPL EGR, a positive differential pressure between the turbine outlet and the compressor inlet is generally needed. However, LPL EGR has slow response than that of HPL systems, especially at low load or speed (Yamashita et al., 2011). Facing the reinforced regulations, exhaust gas recirculation system is widely used and believed to be a very effective method for NOx and PM reductions. Furthermore, increasing needs of low temperature combustion (LTC), EGR have been issued as key technology expecting to provide heavy EGR rate and newly developed dual loop EGR system as the future of EGR types has became a common issue.

The experimental results of dual loop EGR systems were reported in Cho et al. (2008) who studied high efficiency clean combustion (HECC) engine for comparison between HPL, LPL and dual loop EGR at five operating conditions. Adachi et al.(2009) and Kobayashi et al.(2011) reported that The combination of both high boost pressure by turbocharger and a high rate of EGR are effective to reduce BSNOx and PM emissions. Especially, The EGR system using both high-pressure loop EGR and low-pressure loop EGR is also effective to reduce BSNOx and PM emissions because it maintains higher boost pressure than that of the high-pressure loop EGR system alone.

It was also reported that determination of the intake air/exhaust gas fraction by proper control logic (Wang, 2008; Yan & Wang, 2010, 2011) and turbocharger matching (Shutty, 2009) was important. However, there was more complex interaction between variables affecting the total engine system. Therefore, it is necessary to identify dominant variables at specific operating conditions to understand and provide adaptive and optimum control logic.

One of the optimization methods, design of experiment (DOE), can provide the dominant variables which have effects on dependent variables at specified operating conditions. Lee et al.(2006) studied the low pressure EGR optimization by using the DOE in a heavy-duty diesel engine for EURO 5 regulation. The dominant variables that had effects on torque, NOx and EGR rate were EGR valve opening rate, start of injection and injection mass. In their study, the optimized LPL EGR system achieved 75% NOx reduction with 6% increase of BSFC.

In this study, as one of the future EGR types, the dual loop EGR system which had combining features of high pressure loop EGR and low pressure loop EGR was developed and optimized to find the dominant parameter under frequent engine operating conditions by using a commercial engine simulation code and design of experiment (DOE). Results from the simulation are validated with experimental results.

2. ENGINE MODEL

2.1. Engine Specification

The engine specification used to model is summarized in Table 2. An original engine was equipped with a variable geometry turbocharger (VGT), intercooler and HPL EGR system. Operating parameters included engine operating speed, fuel flow rate, ambient conditions, and combustion data. In addition, the length of connecting rods, distance between the piston and pin, compression ratio, and the coefficient of friction were collected and entered in GT-POWER. Data sets of valve diameters, valve timing, injection timing, duration and injection pressure were also acquired. These data were classified information of the engine manufacturer and could not be listed in detail. Engine operating conditions are summarized in Table 3.

The selected 5 operating conditions in the analysis were picked up from frequently operated region at emission test point given by engine manufacturer.

2.2. Engine Analysis Tool

Simulations were carried out by using commercial 1D code, GT-POWER, which is designed for steady-state and transient simulations and can be used for analyses of engine and powertrain control. It is based on one-dimensional gas dynamics, representing the flow and heat transfer in the piping and in the other components of an engine system. The complicated shape of intake and exhaust manifolds were converted from 3D models (by using CATIA originally) to 1D models by 3D-discretizer. Throughout the

2. Engine model

conversion, analysis of gas flow and dynamics could be faster and easier under 1D flow environment.

Table 2. Engine specifications

Item	Specification
Engine volume	3 liter
Cylinder arrangement	6cyl., V- type
Bore, Stroke	84, 89mm
Compression ratio	17.3
Connecting rod length	159 mm
Wrist pin to crank offset	0.5mm
Firing order	1-3-4-2-5-6
Firing intervals	120 CA
Injection type	Common rail
EGR system	High pressure EGR system
Max. torque@rpm	240PS@3800rpm
Max. power@rpm	450N-m@1720~3500rpm

Table 3. Engine operating conditions

Case #	RPM	BMEP (bar)
1	732	2.17
2	1636	4.66
3	1422	3.66
4	1556	9.93
5	1909	8.80

The combustion model was the direct-injection diesel jet (DI jet) model and it was primarily used to predict the burn rate and NOx emission simultaneously.

2.2.1. Overview Of Di Jet

The combustion model, DI jet, was firstly introduced by Hiroyasu known as a multi-zone DI diesel spray combustion (Hiroyasu et al., 1983). The core approach of this model is to track the fuel jet as it breaks into droplets, evaporates, mixes and burns. As such an accurate injection profile is absolutely required to achieve meaningful results. The total injected fuel is broken up into packages (also referred as zones): 5 radial and many axial slices. Each package additionally contains parcels (or subzones) for liquid fuel, unburned vapor fuel and entrained air, and burned gases.

The total mass of fuel in all of the packages will be equal to the specified injection rate (mg/stroke) divided by the specified number of nozzle holes, as DI jet will model the plume from only one nozzle hole.

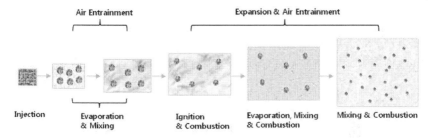

Figure 1. Air-fuel mixing process within each package

The occurring processes in the package are shown in Figure 1. The package, immediately after the fuel injection, involves many fine droplets and a small volume of air. As the package recedes from the nozzle, the air entrains into the package and fuel droplets evaporate. Therefore, the small package consists of liquids fuel, vaporized fuel and air. After a short period of time from the injection, ignition occurs in the gaseous mixture resulting in sudden expansion of the package. Thereafter, the fuel droplets evaporate, and fresh air entrains into the package. Vaporized fuel mixes with fresh air and combustion products and spray continues to burn.

2.2.2. Combustion Process Of Di Jet

Figure 2 shows the detailed combustion process of each package in DI jet model (Hiroyasu et al, 1983). When ignition is occurred, the combustible mixture which is prepared before ignition burns in small increment of time. Combustion rate and amount of burning fuel of each package are calculated by assuming the stoichiometric condition. When the air in the package is enough for burning the vaporized fuel, there are combustion products, liquid fuel and the remained air in the package after ignition occurs. In the next small increment of time, the fuel droplets evaporate and fresh air entrains in the package. The combustion of the next step occurs (case A in figure 2). After this step, since the stoiciometric combustion is assumed, either the vaporized fuel or the air is remained. When the air is remained, the same combustion process is repeated. But when the vaporized fuel is remained, the amount of burning fuel is controlled by the entrained air in the next step (case B in Figure 2). If ignition occurs, but the air in the package is not enough for burning the vaporized fuel, the combustion process continues under the condition shown in case B. Therefore, all the

2. Engine model

combustion processes in each package proceed under one of the conditions shown in Figure 2; Case A is evaporation rate control combustion, and B is entrainment rate control combustion. The heat release rate in the combustion chamber is calculated by summing up the heat release of each package. The pressure and average temperature in the cyinder are then calculated. Since the time histories of temperature, vaporized fuel, air and combustion products in each package are known, the equilibrium concentrations of gas compositions in the package can be calculated. The concentration of NOx is calculated by using the extended Zeldovich mechanism. More detailed governing equations can be found in Hiroyasu's studies (Hiroyasu et al, 1983).

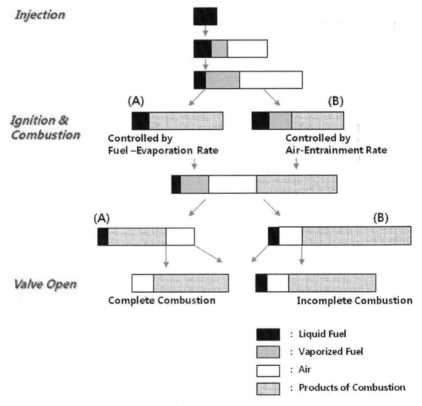

Figure 2. Schematic diagram of the mass system in a package

2.3. Engine Model With Hpl Egr

Based on experimental data, an engine model with HPL EGR was designed. Boost pressure was matched at appropriate turbocharger speeds based on turbine and compressor maps. The injection duration at a given injection timing, injection pressure, combustion pressure and temperature were determined. Then, back pressure at the turbine downstream and EGR valve opening were determined. EGR rate, temperature and pressure drop after the EGR cooler was monitored by installing actuators and sensors. Finally, results of the simulation were compared to the experimental data. Figure 3 shows the Engine model with HPL EGR.

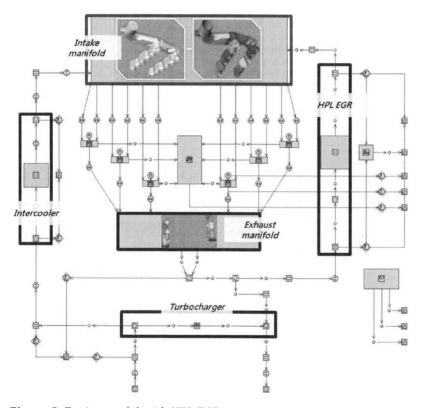

Figure 3. Engine model with HPL EGR

The percent of exhaust gas recirculation (EGR (%)) is defined as following equation.

where [$m_i = m_a + m_f + m_{EGR}$] and m_{EGR} is the mass of EGR and m_a and m_f are the mass of air and fuel.

2. Engine model

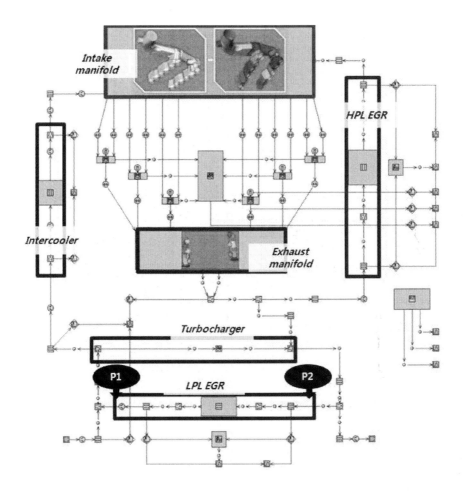

Figure 4. Engine model with dual-loop EGR

2.4. Engine Model With Dual Loop Egr System And Optimization

Based on the HPL model, a dual loop EGR model was designed. Comparing to the HPL EGR system, flap valve opening rate became one of the most important variables for pressure difference at P_2-P_1 inFigure 4.

First, Dual loop EGR simulation was performed under constant boost pressure. Flap valve opening at tail pipe and turbocharger RPM, which had effects on boost pressure and back pressure under dual loop EGR system, were selected as independent variables. In this case, NOx reduction rate would increase, but torque and BSFC would decrease. And the next step, optimization was performed to compensate torque loss and brake specific fuel consumption (BSFC) by modifying injection mass, start of injection (SOI) and EGR valve opening rate. Results of simulation were compared to the HPL and dual loop models in terms of torque, EGR rate, BSNOx, and BSFC.

2.5. Design Of Experiment (Doe)

In this study, optimization based on DOE was performed.

There are main variables which have major effects on torque, BSNOx, BSFC, and EGR rate. In this study, 6 independent variables were selected such as HPL EGR valve opening diameter, LPL EGR valve opening diameter, injection mass, start of injection (SOI), flap valve opening diameter at the tail pipe and turbocharger RPM (TC RPM). Then proper ranges were set and DOE was performed based on the full factorial design.

Torque, EGR rate, BSNOx, BSFC and boost pressure were selected as response variables. The range of each independent variable was chosen based on the engine design performance.

Table 4 shows control factors and levels for optimization of the dual EGR system.

Table 4. Control factors and levels for optimization of the dual EGR system

Control factor		Level 1	Level 2	Level 3
EGR valve	HPL	base	15% open	30% open
	LPL	base	15% open	30% open
Injection mass		base	+2.5%	+5%
SOI		3 CA adv.	1.5 CA adv.	base.
Flap valve		15% close	Base	15% open
TC RPM		-5000 RPM	base	+5000 RPM
Full factorial			$3^6 = 729$	

2.5.1. Control Factors

The 6 independent variables were selected, i.e. HPL EGR valve opening diameter, LPL EGR valve opening diameter, injection mass, start of injection (SOI), flap valve opening diameter at the tail pipe and turbocharger RPM (TC RPM). And their desired ranges are as follows. Base level means the values given by the experimental test.

- HPL & LPL EGR valves: If the EGR valve opens too much, it causes torque loss. In this optimization, the maximum increase of EGR valve diameter was 30% at the given operating conditions from the values under constant boost pressure.
- Injection mass: 5% increase of injection mass was selected and increasing injection mass had an effect on BSNOx. However, it normally degraded BSFC.
- SOI: In general, injection starts faster than 25-23° CA bTDC. If fuel were injected too early, imperfect combustion could degrade the engine performance. In this optimization the maximum advanced CA selected was 3 from the current value which could be within the ranges. Advanced SOI could increase torque without any other variable changes. Also, the EGR rate could increase up to 10%.
- Flap valve and TC RPM: Under dual loop EGR system, pressure difference at P2-P1 in Figure 2was affected by interaction between flap valve and TC RPM which had a dominant effect on EGR rate and BSNOx under the dual loop EGR system.
- Especially, TC RPM was chosen to maintain boost pressure based on the turbocharger map. Positive and negative signs mean increase and decrease of rotation speed of turbocharger shaft which is driven by the exhaust flow. This change in shaft rotation speed is to optimize and maintain target boost pressure under different exhaust energy from combustion at dual-loop EGR system.

3. RESULTS AND DISCUSSION

3.1. Validation

Table 5 shows comparisons between the experiment and the simulation data in terms of injection mass and maximum cylinder pressure. There are two pilot injections and one main injection. By separating pilot injections, combustion noise, soot and NOx can be controlled.

Total injection mass and rate of main injection were given but rate of pilot injections had to be determined by matching injection duration and pressure. Figure 5 shows the torque, EGR rate, and NOx results of the simulation and the experiment, respectively. The differences of each point were within ±5% and it was proven that the simulation results had good agreement with experimental results.

Table 5. Comparison between the experiment and simulation data in terms of injection mass and maximum cylinder pressure

Case No.	Normalized integrated injected mass (fraction)						Maximum cylinder pressure (bar)	
	Experiment			Simulation			Experiment	Simulation
	Pilot 1	Pilot 2	Main	Pilot 1	Pilot 2	Main		
1	0.134	0.134	0.732	0.120	0.130	0.750	53	53
2	0.081	0.081	0.839	0.086	0.089	0.825	58	58
3	0.102	0.102	0.797	0.097	0.122	0.781	51	51
4	0.045	0.045	0.910	0.044	0.035	0.921	75	78
5	0.047	0.047	0.905	0.049	0.045	0.906	76	79

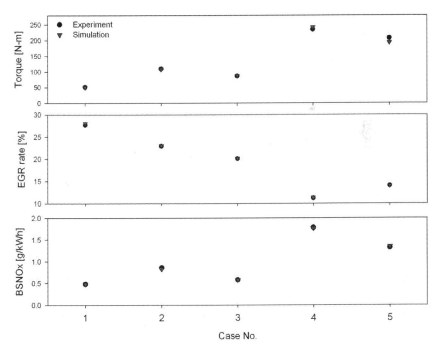

Figure 5. Comparison between experiment and simulation results: Torque, EGR rate and BSNOx

3.2. Optimization Of Dual-Loop Egr System

Based on the DOE, response variables were determined under constant boost pressure with fixed HPL valve diameter. Then, torque and BSFC compensation were performed.

3.2.1. Optimization of dual-loop egr system

Dual loop EGR system optimization was performed based on constant boost pressure and fixed HPL valve opening diameter to minimize torque loss. Table 6 shows the target boost pressure under dual loop EGR system which were from experiment data. Input value of HPL valve opening diameter was the same as that of the HPL model.

Table 6. Target boost pressure

Case No.	1	2	3	4	5
Target boost pressure	1.03	1.23	1.11	1.43	1.56

Figure 6 and 7 show the comparison between combustion characteristics of HPL and dual loop EGR system under constant boost pressure in case 5. Increased EGR rate caused cylinder peak pressure decrease under the dual loop EGR system. And lower peak heat release rate corresponded to lower NOx emissions.

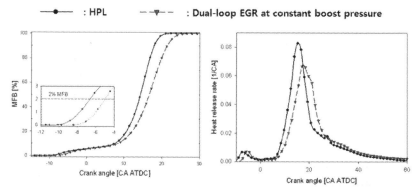

Figure 6. Mass fraction burned and heat release rate between under HPL and dual loop EGR system

Figure 8 shows the result of dual loop EGR simulation under constant boost pressure. Compared to the HPL model, 8% of torque and 8% of BSFC decreased on average. In detail, about 12 % of maximum torque loss (case 3) and about 11 % of maximum BSFC loss (case 1) occurred for the dual loop EGR. On the other hand, about 60% of NOx reduction was achieved on an average. In addition, a maximum of, 80% of NOx reduction was achieved due to the remarkable increase of the EGR rate (case 2). It seemed that the mass of the LPL EGR portion had strong effects on total NOx reduction under larger pressure difference between turbine downstream and compressor upstream. In case 3, the NOx reduction rate became lower because of smaller pressure difference at P2-P1 in Figure 4.

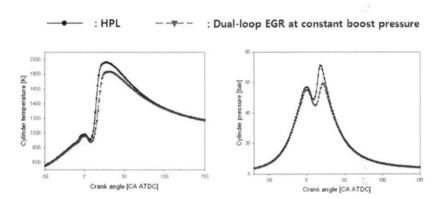

Figure 7. In-cylinder temperature and pressure trace between under HPL and dual loop EGR system

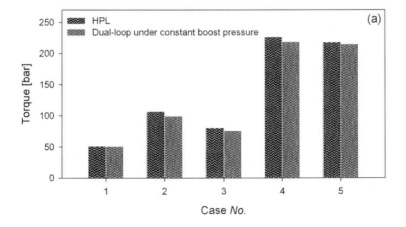

3. Results and discussion

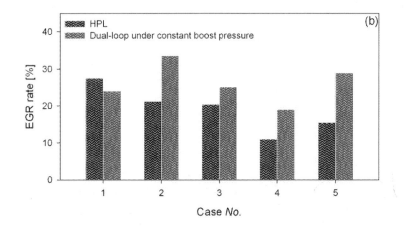

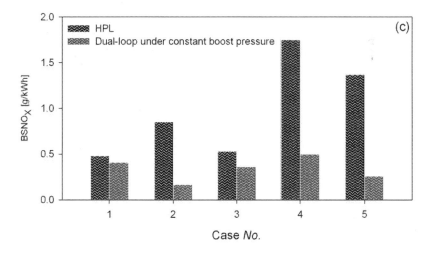

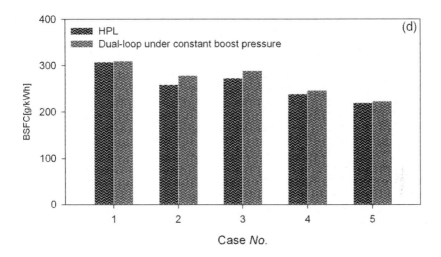

Figure 8. Simulation results comparison between HPL and dual-loop EGR under constant boost pressure; (a) Torque, (b) EGR rate, (c) BSNOx and (d) BSFC

3.2.2. Optimization of dual-loop egr system

To compensate torque and BSFC under constant boost pressure condition, optimization was performed maintaining original boost pressure (bar). By advancing SOI and increasing injection mass, torque and BSFC could be compensated. Table 7 and 8 show results of optimization for constant torque and BSFC with controlled variables.

Figure 9 shows simulation results of HPL and Dual loop EGR under constant boost pressure and optimized Dual Loop EGR, respectively. 8% of torque and 5% of BSFC improvement were achieved on an average compared to the dual loop EGR system under constant boost pressure. Furthermore, higher NOx reduction efficiency appeared at each case except case 4.

3. Results and discussion

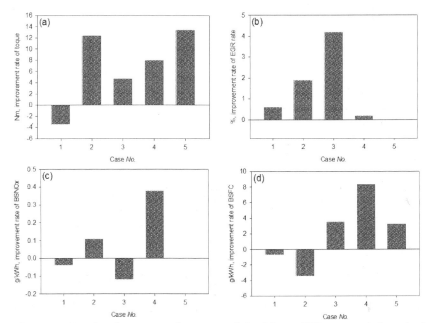

Figure 9. Results deviation of optimized dual-loop EGR compared to dual-loop EGR under constant boost pressure: (a) Torque, (b) EGR rate, (c) BSNOx and (d)BSFC

Table 7. Predicted values

Controlled factors	Case1	Case 2	Case 3	Case 4	Case 5
HPL valve	12% open	3% open	23% open	12% open	7% open
LPL valve	12% open	9 % open	13.5% open	17% open	8% open
Injection mass	+1.5%	+3%	+2.9%	+2.5%	+1%
SOI	-	3 CA adv.	3 CA adv.	3CA adv.	1.2CA adv.
Flap valve	-	-	-	8% open	5% open
TC RPM	+2500	+2500	+5000	+1000	+2000

Table 8. Validation results

	Validation results				
Torque (N-m)	52.24	111.67	82.03	223.87	207.84
BSFC (g/kW-h)	307.2	255.27	276.25	247.01	233.30
EGR rate (%)	24.60	35.48	29.26	19.21	30.13
BSNOx (g/kW-h)	0.37	0.28	0.24	0.88	0.22

In case 4 (1556RPM / BMEP 9.93bar), it seemed that controlled variables which affect torque and BSFC were decoupled with the NOx reduction rate due to relatively high load conditions. It was necessary to control the variables sensitively at high load conditions.

For the optimized dual loop EGR system, 60% improvement of deNOx efficiency was achieved with increasing the EGR rates through all the cases when compared to results for the HPL system.

4. CONCLUSIONS

In this study, engine simulation was carried out to optimize the dual loop EGR system at 5 different engine operating conditions. As a result, the dual loop EGR system in a light-duty diesel engine had the potential to satisfy future emission regulations by controlling the dominant variables at given operating conditions. The details are as follows.

- An engine model for the HPL EGR was developed based on the experimental data at 5 operating conditions. The calibrated simulations showed within ±5% difference with the experimental results.

- Under constant boost pressure conditions, an average 60% NOx reduction was achieved in the dual loop EGR system compared to the results under the HPL system. However, approximately 8% of torque loss and 8% of BSFC loss occurred respectively.

- To compensate torque and fuel consumption, independent variables, such as start of injection and injection mass, were selected as additional control factors. Comparing these variables to the dual loop EGR system under constant boost pressure, approximately 8% of torque and 5% BSFC improvement were achieved except at high load conditions (case 4).

For the optimized dual loop EGR system, 60% improvement of deNOx efficiency was achieved with increasing EGR rate through all the cases compared to results for the HPL system.

REFERENCES

1. T. V Johnson, (2011), Diesel Emissions in Review, SAE International Journal of Engines 2011-01-0304, 4 1 143-157

2. M Zheng, G. T Reader, & J. G Hawley, (2004), Diesel Engine Exhaust Gas Recirculation- A Review on Advanced and Novel Concepts, Energy Conversion & Management, 45 6 (April 2004), 883900 , 0196-8904

3. A Yamashita, H Ohki, T Tomoda, & K Nakatami, (2011), Development of Low Pressure Loop EGR System for Diesel Engines, SAE Technical Paper 2011-01-1413, 0148-7191 Detroit, Michigan, USA, April, 12-14, 2011

4. K Cho, M Han, R. M Wagner, & C. S Sluder, (2008), Mixed-source EGR for Enabling High Efficiency Clean Combustion Mode in a Light-duty Diesel Engine, SAE International Journal of Engines, 1 1 99. 457 EOF465 EOF , 1946-3944

5. T Adachi, Y Aoyagi, M Kobayashi, T Murayama, Y Goto, & H Suzuki, (2009), Effective NOx Reduction in High Boost, Wide Range and High EGR Rate in a Heavy Duty Diesel Engine, SAE Technical Paper 2009-01-1438, 0148-7191 Detroit, Michigan, USA, April, 20-23, 2009

6. M Kobayashi, .; Y Aoyagi, ., T Adachi, ., T Murayama, ., M Hashimoto, ., Y Goto, . & H Suzuki, (2011), Effective BSFC and NOx Reduction on Super Clean Diesel of Heavy Duty Diesel Engine by High Boosting and High EGR Rate, SAE Technical Paper 2011-01-0369, 0148-7191 Detroit, Michigan, USA, April, 12-14, 2011

7. 0967-066112 16 14791486 Wang, J. (2008), Air Fraction for Multiple Combution Mode Diesel Engines with Dual-loop EGR System, Control Engineering Practice, Vol. 16, No. 12, (December 2008), pp. 1479-1486, ISSN 0967-0661

8. F Yan, & J Wang, (2010), In-cylinder Oxygen Mass Fraction Cycle-by-Cycle Estimation via a Lyapunov-based Observer Design, IEEE 2010 American Control Conference, 652657 , 978-1-42447-426-4 Baltimore, Maryland, USA, June 30- July 02, 2010

9. F Yan, & J Wang, (2011), Control of Dual Loop EGR Air-Path Systems for Advanced Combustion Diesel Engines by a Singular Perturbation Methodology, IEEE 2011 American Control Conference, 15611566 , 978-1-45770-080-4 San Francisco, California, USA, June 29- July 01, 2011

10. V. Mueller, R. Christmann, S. Muenz, V. Gheorghiu, 2005System Structure and Controller Concept for an Advanced Turbocharger/EGR System for a Turbocharged Passenger Car Diesel Engine, SAE Paper 2005013888

11. J Shutty, (2009), Control Strategy Optimization for Hybrid EGR Engines, SAE Technical Paper 2009-01-1451, 0148-7191 Detroit, Michigan, USA, April, 20-23, 2009

12. S. J Lee, K. S Lee, S Song, & K. M Chun, (2006), Low Pressure Loop EGR System Analysis Using Simulation and Experimental Investigation in Heavy-duty Diesel Engine, International Journal of Automotive Technology, 7 6 (October 2006), 659666 , 1229-9138

13. J Park, K. S Lee, S Song, & K. M Chun, (2010), A Numerical Study for Light-duty Diesel Engine with Dual Loop EGR System under Frequent Engine Operating Conditions by Using DOE, International Journal of Automotive Technology, 11 5 (October 2010), 617623 , 1229-9138

14. H Hiroyasu, T Kadota, & M Arai, (1983), Development and Use of a Spray Combustion Modeling to Predict Diesel Engine Efficiency and Pollutant Emissions: Part 1 Combustion Modeling, Bulletin of the JSME, 26 214 (April 1983), 569575 , 0021-3764

Index

A
automobiles, 43, 102, 130, 147, 162
automotive, 71, 72, 74, 91, 96, 99, 135, 165, 167, 277

C
carbonaceous, 37, 48, 49, 58, 65, 66, 130
classification, 71, 72, 73, 78, 92

D
diagnosis, 71, 72, 73, 90, 91, 92, 303
diesel engine, 2, 3, 4, 5, 6, 7, 14, 231, 232, 233, 238, 273, 275, 276, 277, 279, 293

L
Legislation, 130

O
oxidizing species, 131

S
screw expander, 209, 210, 211, 216, 217, 219, 221, 222, 224, 231, 235

T
temperature, 4, 7, 11, 19, 20, 21, 224, 233, 234, 238, 239, 245, 277, 278, 282, 283, 289
turbocharger, 110, 173, 174, 175, 176, 177, 179, 186, 187, 285, 286

Z
zero-dimensional model, 174, 176, 177, 190, 202